BIOLOGY
PRACTICAL MANUAL

BIOLOGY
PRACTICAL MANUAL

By
Dr. S. Sundara Rajan
Prof. and Head, Dept. of Botany
St. Joseph's College
Bangalore - 560 025

ANMOL PUBLICATIONS PVT. LTD.
NEW DELHI - 110 002 (INDIA)

ANMOL PUBLICATIONS PVT. LTD.
4374/4B, Ansari Road, Daryaganj
New Delhi - 110 002
www.anmolbooks.com

Biology Practical Manual

First Edition, 2001

ISBN 81-261-0989-0

PRINTED IN INDIA

Published by J.L. Kumar for Anmol Publications Pvt. Ltd., New Delhi and Printed at Mehra Offset Press, Delhi.

CONTENTS

PREFACE

Laboratory studies constitute an integral part of science study. It is in the laboratory that the students come into contact with the objects about which they would study in theory classes. It is in the laboratory again the students conduct experiments and convince themselves about the veracity of various theoretical concepts.

In Biology, laboratory study of specimens, materials etc., provide the students first hand knowledge and help them in proper comphrehension.

Practical Manual of Biology has a direct approach and deals with the subject matter purely from the laboratory point of view. Theoretical details and concepts which are superflous for a practical study are dispensed with.

The manual has 24 chapters dealing with all aspects of biological study—morphological, physiological, anatomical and biochemical. Experiments, structural details etc. are explained in simple language to make the students' understanding easy.

Criticism and suggestions from the actual users of the book are most welcome.

I am thankful to Mr. J.L. Kumar, Managing Director, Anmol Publications Pvt. Ltd., New Delhi, for publishing this book.

—**Dr. S. Sundara Rajan**

St. Joesph's College
Bangalore - 560 025

CHAPTER 1

INTRODUCTION

The study of science subjects is carried out at two levels,viz., *Theoretical* and *Practical.* Theoretical study is carried out in lecture classes when the students are imparted all the details regarding a particular topic. This has to be supplemented by practical or Laboratory studies, when the students come into active contact with the objects about which they would have heard in lecture classes. Therefore it may be safely said that laboratory studies form an integral part of a comprehensive study of any topic in science subjects.

In the study of plants, laboratory assumes a special significance as the students acquire a meaningful and firsthand knowledge about the plants, animals which they study in lecture classes. In fact as far biology is considered any study of a plant, animal without being supplemented by laboratory observation is futile. It is only in the laboratory study that the student can dissect, observe and analyze the various features of the plant,animal and it is only a study of this type that makes the student acquire a meaningful knowledge about the organisms. Besides observation and study in the laboratory plants are also to be studied in field to obtain an account of their habitat, habit, association and such other details.

In this chapter an attempt is made to explain some of the essentials that a student should understand before he begins a practical work.

Cleanliness: This is the most important point that a student should keep in mind while in the laboratory.

The working table and the surroundings should be kept dry and spotlessley clean.. All the requirements needed by the students such as pewter dishes, slides, reagent bottles, stains etc., must be properly labelled and arranged neatly. The microscope and particularly the optical parts must be cleaned with a white lint free cloth. After the completion of the assigned work, the student should take care to clean the place, remove any debris of the material and put it into the dust bin.

The students should work in silence and independently unless the assigned experiment is to be carried out by a group. When a particular work is assigned to a student, he must carefully plan and divide the available time for experimentation, observation and recording of the results. Whenever a doubt arises, it is always better to consult the teacher in charge, than a classmate. It is only by a proper and systematic work, that a student can derive maximum benefit from the practical classes.

Laboratory requirements

The department provides certain amenities, instruments etc. for each student for the purpose of practical work. But a student also should carry some of the requirements in order to be effective in a practical class. One cardinal principle that every student should remember in a practical class is to have one's own requirements and not to depend on anyone. Never lend or borrow any requirements. Borrowing or lending a pencil, needle or blade will disturb the work of both.

In general every student should individually possess the following requirements

1. *Two sharp pointed needles* mounted on stainless steel or plastic handle. This is essential for the purpose of separating and teasing the microscopic material.
2. *A pair of forceps:* This is essential to remove the material either from the bottle or from pewter dish.
3. A sharp properly honed razor to cut sections of the plant material.
4. A new safety razor blade. This will help either to cut section or to trim the material.
5. Two fine hair paint brushes. This will be of use in lifting the sections of plant (with out damage) from the pewter dish on to the slide.
6. Two sharpened pencils (H, Hb or 2b).
7. A pencil eraser.
8. A pencil sharpener.
9. Glass droppers.
10. A piece of thin, white, lint free cloth to clean the optical parts of the microscope.
11. An observation book to note down immediately the points of interest.
12. A practical record.
13. Glass slides and cover slips(if they are not provided by the department).

Scheme of Practical Work

Practical work in the laboratory will be most useful and effective if the student keeps in mind the following:

1. Maintain silence during working. All the requirements for the practical work must be absolutely clean.
2. Every student should work independently, unless the experiment involves the participation of others.
3. Before beginning the practical work, listen to the instructions of the teacher attentively and work accordingly.
4. Follow a systematic procedure in a study.
5. If the work involved is the morphological study of an organism, follow the procedure given below:
   (a) Study of external characters - for example whether a plant is microscopic or macroscopic, unicellular or multicellular, filamentous or colonial etc.
   (b) Study of internal structure like details of cell tissues etc. If the material requires sectioning, sections are to be cut and internal details should be observed.
   (c) Study the reproductive details like asexual and sexual reproductive organs.

1. Mount the material after properly teasing and separating the filaments, mycelia etc. Stain the material wherever necessary and cover the preparation with a cover glass.
2. Observe the slide carefully under the microscope, first under low power and then under high power.
3. Make sure that you have observed all the details as instructed by the teacher.

After all the observations are over, record your findings, structural details etc. in the practical records in the form of diagrams. Draw proportionate diagrams in the proper place on the drawing sheet. Diagrams should never be drawn to the very edge of the drawing sheet. Write suitable explanatory notes pertaining to the diagrams on the interleaf.

Laboratory Equipments

In addition to the individual requirements carried by the student, the department generally provides the following equipments/instruments etc.to the students. Sometimes these equipments may be given to each student.

These are :

1. Student's compound microscope .
2. Dissection microscope.
3. Watch glasses.
4. Binocular Research microscope.
5. Cavity blocks.
6. Slides and cover glasses.
7. Pewter dishes.
8. Enamel tray.
9 Polythene/glass wash bottle.
10. Spirit lamps.
11. Glass marking pencil.
12. Microtome.
13. Staining /bottles.
14. Stone/ for honing the razor.
15. Chemicals and stains.
16. Culture media(to be prepared by students).

In the list of requirements mentioned above, we will discuss some in detail.

Microscope

Of all the instruments used by students in the biology laboratory, the one that is most indispensable to the students of biology is the 'Microscope'. It is not exaggeration to say that no biological study in the laboratory is possible without the microscope.

The function of the microscope is to enlarge the small objects, so that they become visible to the human eye. In reality the magnification power of the microscope lies in its resolving capacity between one particle and another. A microscope distinguishes one particle from another which are situated very closely. A human eye cannot resolve particles that are lying closer than 0.025 mm. In its function a microscope can be compared to a human eye, but much more effective than that.

There are different types of microscopes, but all of them are based on the same principle with variations in their construction. Some of the types of microscopes used in the laboratory are:

(1) Compound microscope.
(2) Binocular Research microscope.
(3) Dissection microscope.
(4) Stereoscopic dissection microscope.
(5) Phase contrast microscope.
(6) Electron microscope etc.

Of the six types of microscopes mentioned above the compound microscope and the dissection microscope are the ones most frequently used by students. The Electron microscope is very costly and a very highly sophisticated instrument.

In this book the compound microscope and the dissection microscope are explained in great detail as the students should be familiar with them.

Compound Microscope

The microscope has a strong 'U' shaped or 'V' shaped base. It is also called the horse shoe shaped base. Attached to the base is a bent limb. This limb can be tilted in any direction as it is attached to the base with an inclination joint.

Attached to the limb is a stage and a body tube. The body tube is attached to the limb with the help of a rack and pinion so that it can move up and down. To help in the movement of the tube during focusing, there are 2 knobs. Viz.,coarse adjustment knob and fine adjustment knob.

The body tube at its top end consists of the ocular or eye piece lens. The eye pieces usually have three magnifications 5x, 10x,and 15x, and can be used to get a low or high magnification. Attached to the lower end of the tube is a revolving nose piece with three objective lenses low (10x) medium (40x) and high (60x). Some times there will be a fourth objective of the magnification of 100x. This is also called the oil immersion lens (while viewing through an oil immersion lens a drop of oil-liquid paraffin has to be put on the viewing slide).

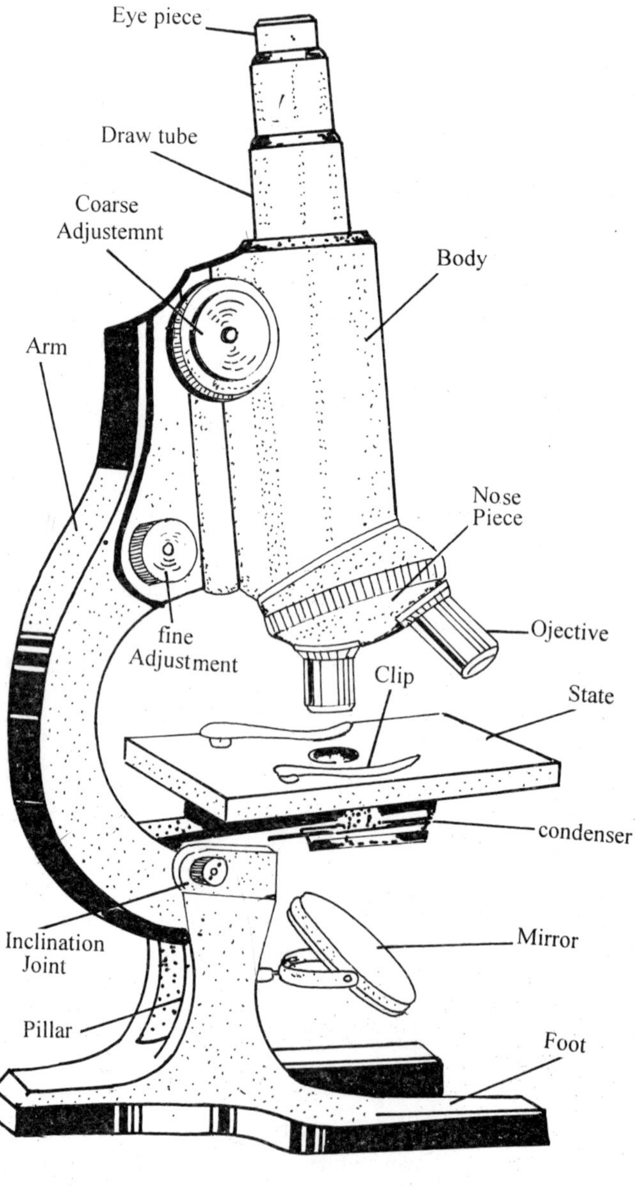

Fig. 1.1 Parts of a compound microscope

Below the objectives and attached to the limb is the *stage* on which the slides are kept for viewing. The stage may be circular, square or rectangular in shape. In order to keep the slides properly placed on the stage, there are two stage clips. In some of the costly microscopes, the stage has movable sliding system operated by knobs to move the slides. This is called the mechanical stage. The mechanical stage permits the movement of the slide backwards or forwards, left or right with the turning of the knobs provided for the purpose. A hole is provided in the stage to receive the light from below and illuminate the object. Attached to the limb, below the stage is a condenser (fixed or movable up and down) lens. Below the condenser lens is an iris diaphragm to regulate the light passing through. Iris diaphragm is a metallic disc with an operator. The operator can be narrowed or widened at the turn of a lever. The condenser which is a system of lens condenses the light rays passing through and converges them on the object.

The source of illumination in a student's microscope is a mirror which reflects the natural light or electric light. The mirror has two faces - plain and concave and is attached to the limb. The mirror is adjustable and can be turned to the direction of light. When day light is the source of illumination, the concave side of the mirror is usually used to converge the rays of light. In some of the costly microscopes, instead of the mirror there is a 'built in illumination'. This is a tube with a lamp (9v or 12v) attached to a transformer, which can be plugged to any AC source.

Operation of the Microscope

Microscope should be placed in a well illuminated area but avoiding the direct light as it is harmful to the eyes. Generally northern light is preferable. When ' built in illumination' is employed, it is always better to use a blue or green filter to subdue the light.

The light has to be first focused on the condenser by adjusting the mirror towards the source of light. The amount of light entering the condenser may be regulated by closing or opening the iris diaphragm. The slide to be viewed is kept on the stage and fixed securely with the help of the stage clips. The object is first observed with a low power objective and then turned on to medium or high power. While viewing under high power the specimen or section has to be covered with a cover slip. The slide should be observed with both the eyes open even while observing with monocular microscope.

Care of the Microscope

1. Clean all the parts of the microscope before and after use. The lenses and the mirror are to be cleaned with a lint free cloth or tissue paper.
2. Do not pull the microscope to the very edge of the table.
3. While lifting the microscope use both hands with one hand holding the base.
4. The slide to be observed must be clean and dry (specially the under surface of the slide).
5. Do not remove the eyes pieces or objectives from the microscope.
6. The microscope should be covered when not in use.

Binocular Research Microscope

This is like the ordinary compound microscope except that, it has binocular head for providing a natural stereoscopic view.

Dissection Microscope

This is a very simple microscope with only one set of lenses. Microscopes of this type will not give a very high magnification, but are very useful in dissecting flowers, plant parts etc.

The microscope has a 'V' shaped base and a limb. The stage is made up of a transparent glass plate. Below the stage is a mirror to focus the light. Attached to the limb is a mechanical arm which can be moved up and down with the help of a rack and pinion movement. The arm can also be moved horizontally. Attached to the free end of the arm is a holder to accommodate the lens (10x,15x etc.).

Operation : The specimen to be observed / dissected is placed on the stage and located with the help. Floral parts or others may be then dissected with the help of two needles, while still observing the specimen.

Stereoscopic Binocular Dissection Microscope

This is a combination of compound microscope and dissection microscope. While it has the higher magnification power of a compound microscope with two sets of lenses, it has retained the versatility and the manoeuvourablity of a dissection microscope.

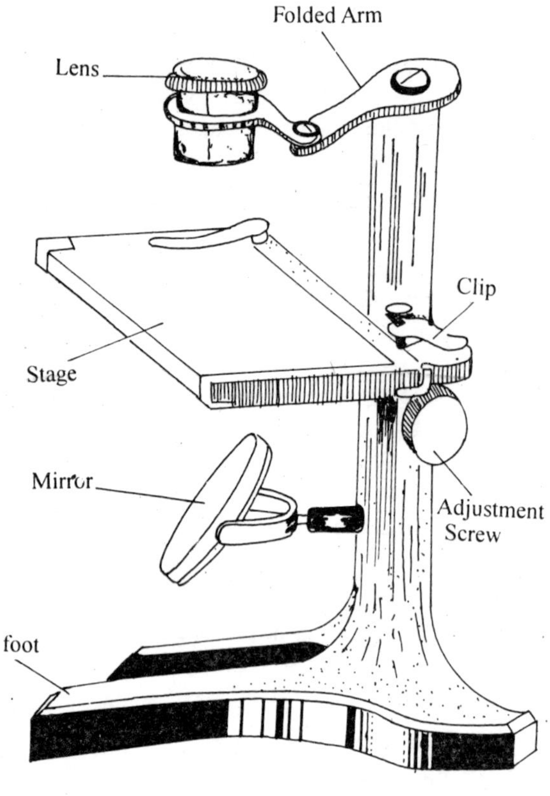

Fig. 1.2 Parts of a dissection Microscope

Phase contrast and Electron microscope are not normally used by students.

Record of Practical work

Practical work carried out in the laboratory should be recorded. The biology practical record book has two types of sheets viz., Interleaves and drawing sheets. Interleaves are meant to write the notes pertaining to the diagrams. The drawing sheets which are thicker than the interleaves are used for diagrams. The following points may be useful in maintaining a good record.

1. Use good pencils (H and HB) with sharp points and a good quality eraser.
2. Leave a minimum of one inch margin on all the sides.
3. Always write diagrams of only one organism in a page. Do not mix up diagrams of other plants.
4. On the left hand corner of the drawing sheet write the date of the experiment.
5. Give the complete systematic position of the member on the top of the drawing sheet.
6. For microscopic organisms draw first the habit sketch followed by low power and high power views. High power view should show the cellular details.
7. In th case of anatomical drawings draw the ground plant first, followed by a sector enlargement to show the cellular details.
8. All parts of the diagrams must be properly labelled.
9. Stages of reproduction if any must be drawn after the details of vegetative structure.
10. The description corresponding to the diagrams must be written on the opposite interleaf.

CHAPTER 2

STUDY OF CELL

STUDY OF CELL

Purpose : To study the structure of a typical plant cell under microscope by using thin layer of onion peel.

Materials : Onion bulbs, slides, coverslip, microscope and stain (saffranin).

Procedure : Carefully peel off a thin layer from the bulb of the onion. Initially it may be difficult to get a thin layer but after the practice it will be possible to get a thin layer which is suitable for microscope examination. For this purpose carefully separate one fleshy leaf of the onion bulb. Bend it in the middle so that it incompletely breaks. When you separate the two pieces a thin layer will come out. Immediately after removing place the onion peel in a petridish containing water.

Cut a small piece of the onion peel suitable for observation and place it in a drop of water in the centre of a clean glass slide. Add a drop of stain, put a cover glass and observe under microscope.

Observation and conclusions. The study shows the following structural details of the plant cell.

1. A number of rectangular cells can be seen under the field of the microscope.

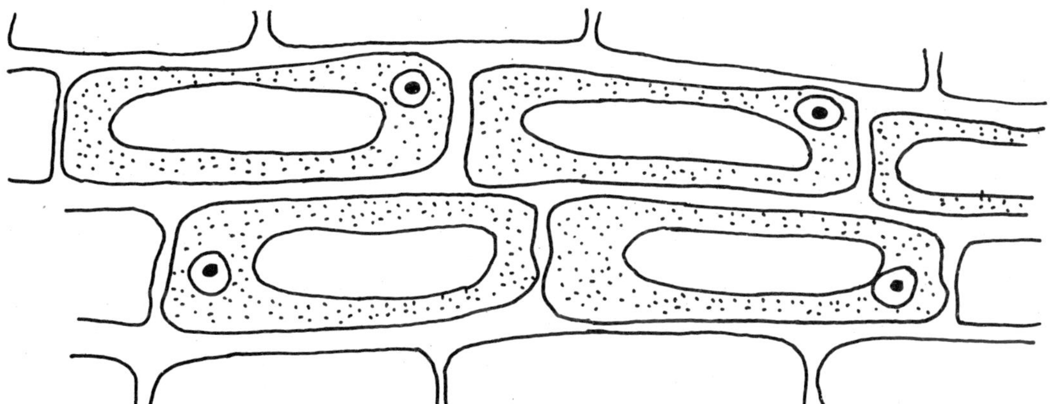

Fig. 2.1 Few cells of onion peel to show cell structure

2. The cells are many times longer than broad and are joined end to end and side to side to form a layer.
3. The cell wall which is the outermost layer can be clearly demarkated.
4. The cell membrane is not very clearly visible.
5. Occupying the centre of the cell is a large vacuole.
6. Cytoplasm is peripheral and occupies the space between the cell membrane and the vacuole.
7. Lying in the cytoplasm is a large vacuole which is red coloured (because it has taken saffranina stain).
8. Also distributed in the cytoplasm are seen a number of leucoplasts.

PROTOPLASMIC STREAMING

Materials : Leaves of Elodea, staminal hairs of Tradescantia, slides, cover slips, water, compound microscope.

Procedure : Collect fresh leaves or dissect out the staminal hairs. Place them on the slide and cover with the help of cover slip adding a drop of water. Observe under high paver of the microscope, specially the chloroplasts in Elodea leaf.

Observation : The chloroplasts appear to move in a circulatory fashion around the vacuole.

In the case of Elodea there is a large central vacuole and the chlorplasts move in a clockwise or anticolock wise direction. This is known as Rotation. In the staminal hairs of Tradescantia there are many vacuoles and the contests move around each vacuole independently and it is known as Circulation.

Inference : The movement of the chloroplasts seen under the microscope is actually due to the circulation or streaming of protoplasm. The movement probably occurs due to the sol-gel transformation of protoplasm. Microtubules of cytoplasm may also contribute to the movement.

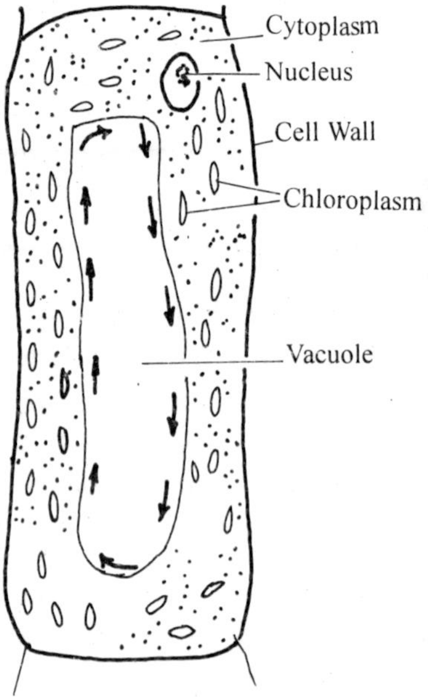

Fig. 2.2 A cell of Elodea leaf showing cyclosis

STUDY OF BACTERIA (Gram's staining)

Materials :

1. Milk sample and cultures of *Staphylococcus* and *Escherichia coli*
2. Gram's staining reagents
   a. Iodine solution
   b. Ethyl alcohol (95%)
   c. Safranin
3. Staining tray, wash bottle, droppers
4. Distilled water
5. Glass slides
6. Inoculating loops
7. Bunsen burner / spirit lamp
8. Microscope

Procedure :

To study bacteria present in milk (Lactate bacteria)

1. Take a clean and dry slide.
2. Put a drop of milk and gently spread to make a thin smear; allow it to dry.
3. Stain the smear with crystal violet stain.
4. Follow the procedure given below for *E.coli* and *S.aureus*

To study E.coli and S.aureus

1. Prepare separate smears of *E.coli* and *S.aureus* on two different slides. Fix these preparations after allowing them to dry. Heating may be done lightly.
2. Stain the smears with crystal violet stain for about 30 seconds.
3. Rinse with water for a few seconds.
4. Add few drops iodine solution to both the smears. Allow the solution the react for 30 seconds.
5. Rinse with distilled water.
6. Add few drops of 95% ethyl alcohol to the smears. Add the alcohol drop by drop and stop adding alcohol when no more colour flows out from the smears. For a thin smear 10-20 seconds time may be enough for complete decolourization of gram negative bacteria.
7. Counter stain both the smears with safranin for about 20-30 seconds.
8. Wash with distilled water and blot dry using an absorbent paper.
9. Air dry the slides.
10. Add a drop of Canada balsam (permanent preparations) or glycerin (temporary mounts), put a cover glass each over the preparations and observe under the oil immersion objective of the microscope.

Observations:

1. Indentify which smear has taken the violet stain and which has taken only the counter stain.
2. Make sketches and classify them based on gram reaction (positive or negative).

Conclusions : Bacteria that appear blue are referred to as gram positive and those appearing pink are called gram negative.

Staphylococcus aureus bacteria appear purple or blue in colour hence they are classified as gram positive while *E.coli* bacterial rods appear pink hence they are classified as gram negative.

In the milk smear three kinds of bacteria are seen. These are *Lactobacill* which are rod shaped gram positive bacteria, they appear purpie or blue; *Streptococcus* - chains of circular cells which are gram positive and appear purple or blue and *Micrococcus* - very small, spherical gram positive bacteria which also appear purple or blue.

CHAPTER 3

PLANT MORPHOLOGY I

CYANOBACTERIA
NOSTOCALES
NOSTOCACEAE
NOSTOC

Common species: *N. calcicala, N. commune, N. sphaericum*

Distribution: Fresh water ponds and ditches. Form mucilagenous balls. Often found on wet rocks, also found in symbiotic association with *Anthoceros, Cycas* and *Lichens*.

Study: 1 External features

2. Reproductive structures
3. Systematic position and identification

External features

1. Many hundreds of filaments are enclosed in the form of a mucilage ball.
2. The ball like colony has a thick mucilage envelop and may be coloured.
3. The plant body is a filament and is enveloped by a mucilage sheath.
4. The filaments are unbranched and do not have a base or apex.
5. Each filament is made up of a number of bead like cells joined end to end.
6. Here and there in the filament certain thick walled cells are seen. These are called heterocysts and may be intercalary or terminal in position.
7. Heterocysts are thick walled and pale yellow coloured. Terminal heterocyst has a single polar module, while intercalary heterocysts have two polar nodules.
8. Vegetative cells have a typical cyanophycean structure. Protoplast having centroplasm and chromoplasm. Cell wall is cellulosic.

Reproductive features

1. Sexual reproduction has not been reported.
2. Asexual reproduction is by the formation of hormogones and akinetes.
3. Hormogones are fragments of the filament, that generally separate from the main filament near a heterocyst.
4. Hormogones develop into new individuals after separation from the main filament.

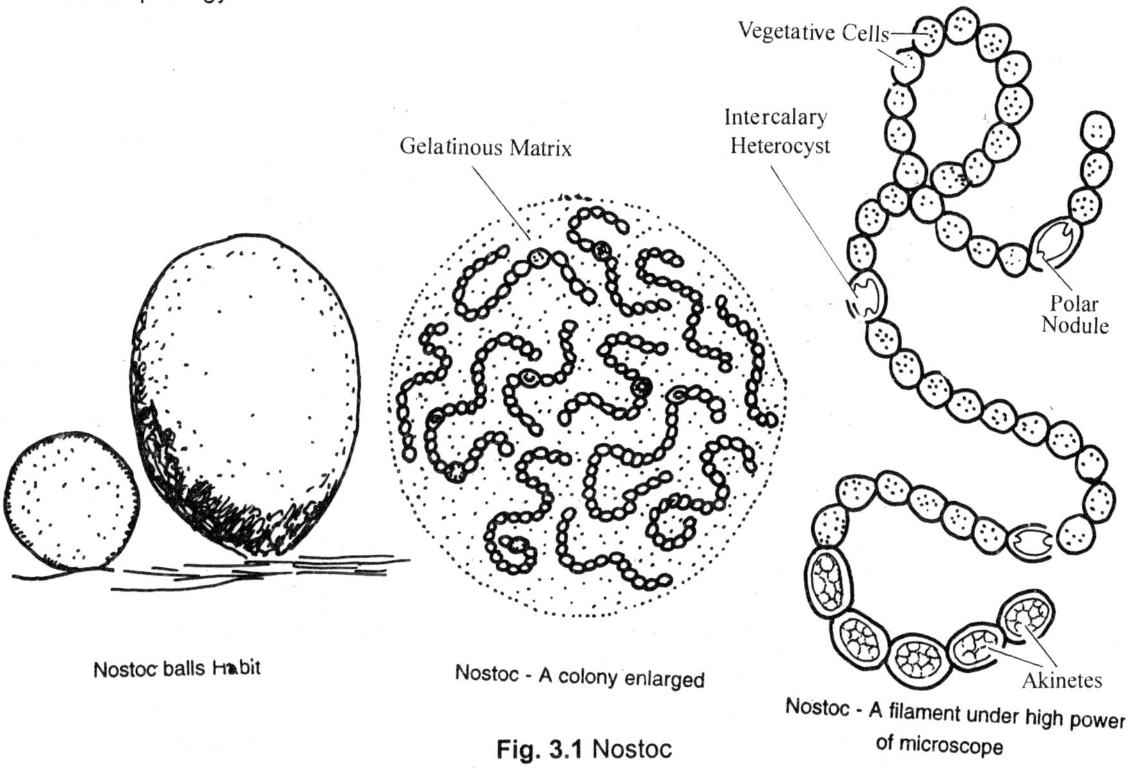

Fig. 3.1 Nostoc

A. Habit, **B.** Colony enlarged, **C.** Filament under high power of microscope.

5. Akinetes are thick walled cells rich in reserve food materials. They separate from the plant and develop into a new individual.
6. Occasionally heterocysts (N. commune) may also germinate to form new individuals.

The Bacillariophyceae are commonly called diatoms. They include a large number of unicellular and clonial algae, which very greatly in form and size. They are most widely distributed among algae. The primary distinguishing feature of Bacillariophyceae is that the cell wall is silicified and consists of two halves, which fit together as do the lid and box. The cell has one to many chromatophores.

Reproduction is by cell division and auxospore formation. Auxospores are usually zygotic in nature, produced by the union of gametes, but may not be always so.

STUDY OF DIATOMS

Common species: P. major, P. viridis, etc.

Distribution: Commonly found in fresh water ponds, and ditches; forms on oily layer on top of water.

Study

1. Cell structure
   (a) Girdle view
   (b) Valve view
2. Reproduction
3. Systematic position and Identification

Cell structure:

1. Structurally the diatom cell consists of two parts or halves which fit in like a lid and box.
2. Of the two halves, the smaller one is called *hypotheca* and the bigger one is called *Epitheca.*
3. The cell has a size of 140 M x 20 M
4. Each valve or half consists of a flat top (valve face) and the margin or girdle representing the overlapping face.
5. A diatom cell can be viewed from two different angles. The view from the side is called girdle viev', and the view from the top is called the valve view.

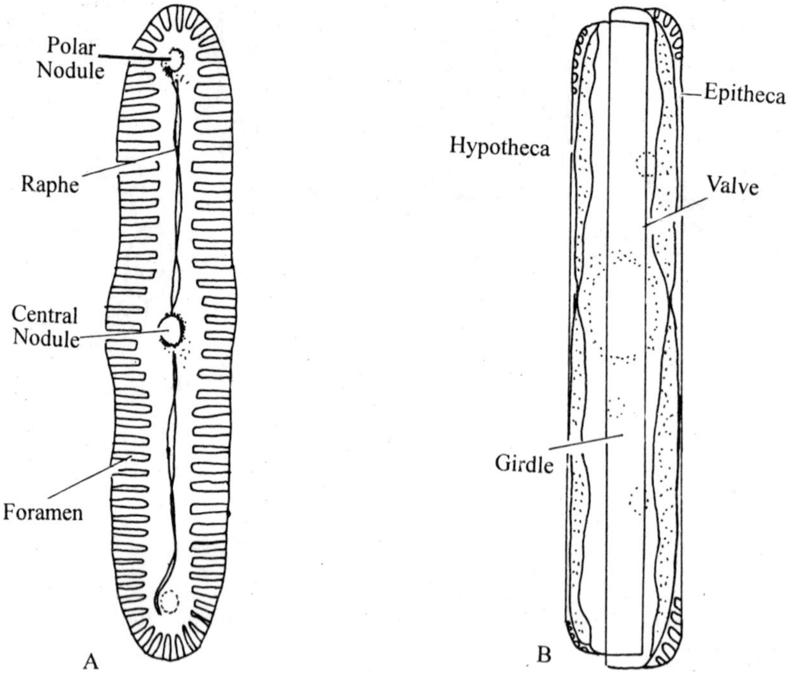

Fig. 3.2 Diatoms **A.** Valve View, **B.** Girdle View

Valve view:

1. Only one valve can be seen.
2. The wall can be clearly seen and it is called frustule
3. The cell wall is made up of pectin and it is silicified.
4. On the wall are seen prominent markings called costae. The costac represent raws (striae) of small pores or cavities in the frustule.
5. Along the centre of the vauve is present a long slit in the wall. This is called Paphe, it is sigmoid in shape.
6. At the centre of the raphe is a swelling called the *central nodule.*
7. At either end of the raphe are also present two nodules called *Polar nodules*
8. The mucilage extends out of the wall through poenings or pores and is supoosed to help in locomotion.
9. In side the silicified wall is the cytoplasmic lining, surrounding vacuole.

Girdle view:

1. In the girdle or side view both the valves are seen (only one valve is seen in the valve view)
2. Costae can be seen in the frustule.
3. Inside the wall is the cytoplasm having a single nucleus.
4. One or two chromatophores may be seen distributed in the cytoplasm.
5. Oil and volution globules are present in the cytoplasm

Reproduction

1. The commonest method of reproduction seen in diatoms is cell division.
2. During cell division, the two halves get separated after the nuclear division. Each daughter cell gets only one valve from the parent. It has to develop another valve by itself. The newly formed valve is always a hyopthece. This is irrespective of the fact, whether it gets a hypotheca or epitheca from the parent cell.

Auxospores:

1. Auxospores are alongated spores formed in the diatom cell to bring about reproduction.
2. Auxosopres are both asexual as well as sexual. They are formed by any one of the following five ways.
   (i) Two cells conjugate to form a single auxospores.
   (ii) Two cells conjugate to form two auxospores
   (iii) Two cells being enveloped into a common envelop, but each giving rise to an auxospores without conjugation
   (iv) One cell gives rise to one one auxospores.
   (v) One cell gives rise to two auxospores.

The first two methods given above are obviously sexual. The last three methods appear to be asexual.

STUDY OF YEASTS

Common species

S.cerveisiae, S.ludiwigii, Schizosaccharomyces octosporus

Occurrence : The members are commonly called yeasts. They are widely distributed in nature. Yeasts are saprophytic and are commonly found in substrata rich in sugars like nectar of flowers, surface of fruits like grapes apples etc. Milk and other food undergo fermentation due to yeasts.

Study :

1. Method of culturing yeasts
2. Vegetative structure
3. Reproduction
4. Systematic position and identification

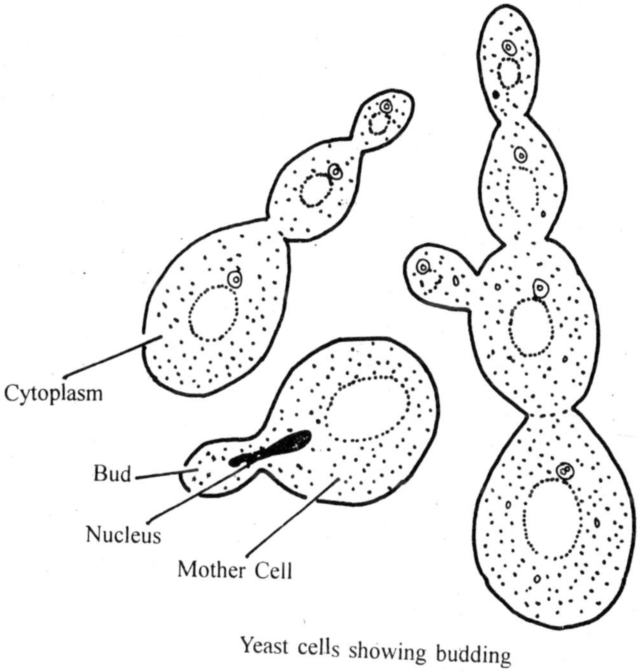

Fig. 3.3 Yeast cells showing budding

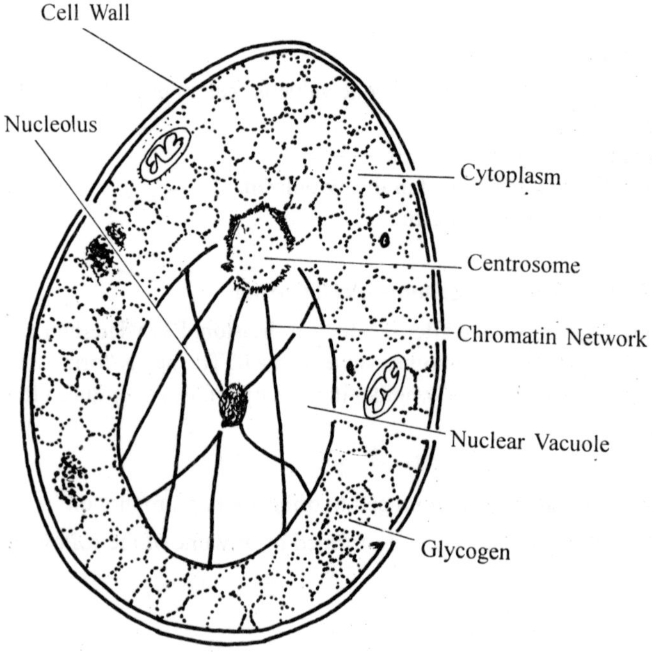

Fig. 3.4 A highly magnified cell of Saccharomyces

Culture of yeasts

1. Place small pieces of yeast cake in sugar solution

2. Yeast cells multiply and a dense cluster of cells is produced

3. After few days the yeast cells can be harvested

Vegetative structure

1. The thallus of yeast is made up of isolated cells (unicellular plant body)

2. The cells are usually oval or spherical ranging from 5-10m in diameter

3. Each cell has an outermost cell wall made up of chitin. Cellulose is absent

4. The cytoplasm shows differentiation into outer ectoplasm and inner endoplasm
5. There is a large vacuole in the center
6. Placed immediately above the vacuole is the single nucleus (the nucleus was mistaken previously as the centrosome)
7. Cytoplasm is granular and contains reserve food materials like glycogen and oil
8. Mitochondira and starch grains are also found in the cytoplasm

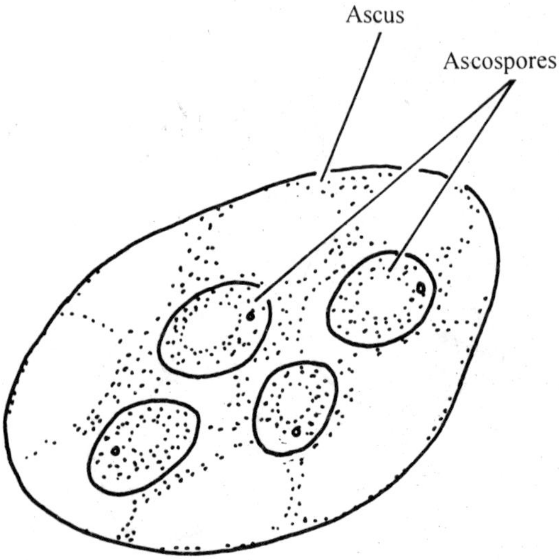

Fig. 3.5 Single ascus of Saccharomyces

Asexual reproduction

1. Yeast cells reproduce asexually by two methods viz., *fission* and *budding.*

2. Fission yeasts are called *schizosaccharomyces* while budding yeasts are called *saccharomyces.*

3. During fission, the single nucleus gets elongated.

4. A constriction appears in the center of the nucleus and it becomes dumb-bell shaped.

5. With the deepening of the constriction, the nucleus gets divided into two bits.

6. The two daughter nucleu migrate to opposite poles. A cross septum appears in the center of the cell dividing it into two.

7. The two daughter cells formed thus grow into adults cells.

8. Budding is another method of reproduction. During budding a small bulge arises from one side of the cell. The bulge increases in size and is now called a bud .

9. Meanwhile, the nucleus of the mother cell divides into two daughter nuclei; one of those migrates into the bud.

10. A small constriction appears at the region of attachment of bud and the parent cell.

11. The constriction deepens and the bud gets separated. It grows into a new individual.

12. In some cases a bud does not get separated from the mother cell but, produces another bud. In this way a chain of buds is formed. This chain may be branched or unbranched. The chain looks like a mycelium hence it is called a Rhizomycelium or a Pseudomycelium.

13. Eventually the chain breaks and the individual cells get separated and grow into new cells.

Sexual reproduction

1. Sexual reproduction takes place by means of gametangial copulation, with two cells functioning like gametangia.
2. The two gametangia fuse forming a diploid zygote.
3. The zygote which is diploid functions like an ascus.
4. The diploid nucleus is ascus undergoes reduction division followed by a mitosis to form eight ascospores. In some cases only four ascospores are formed.
5. Ascospores on liberation develop into new individuals.

Diagrams to be drawn

1. Yeast cell magnified.
2. Yeast cells showing fission and budding.
3. Ascus with ascospores.

Systematic position and identification :

A. 1. Chlorophyll absent

2. Reserve food is glycogen ....**Fungi.**

B. 1. Mycelium is septate or plant body unicellular

2. Characteristic spores - ascospores are produced inside an ascus.

3. Ascospores usually light, rarely otherwise......................**Ascomycetes.**

C. 1. No fruiting body.

2. Saprophytic nutrition

3. Asci not in clusters ....**Endomycetales**

D. 1. Asexual reproduction by budding or fission

2. Plant body unicellular ....**Saccharomycetaceae**

E 1. Members produce enzyme zymase and bring about fermentation.

2. Plant body unicellular ....**Saccharomyces.**

STUDY OF AGARICUS

Common species : *A.campestris bisporous*

Occurrence : Commonly called mushroom, the genus Agarious is a widely distributed saporphytic, fungus growing on dead logs, timber on soil with plant of leaf let.

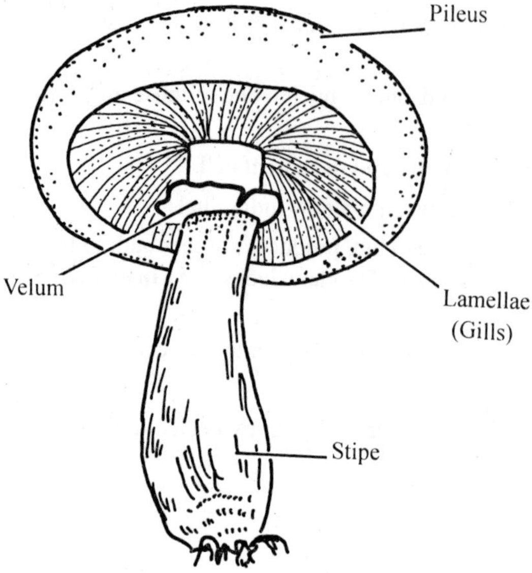

Fig. 3.6 Agaricus– A basidiocarp

Study : 1. The vegetative mycelium
2. Basidiocarp
3. Structure the gill
4. Systematic position and identification

Vegetative mycelium

1. The actula plant body of Agaricus is a subterranean mycelium.

2. The mycelium is extensively branched with anastomosing hyphae. The network of hyphae are called spawns.

3. The hyphae are septate tubular and hyaline.

4. The cyliplasm is granular containing oil globular and other serve food materials.

5. The mycelium is perennial and survives for several years producing basidiocarps which come above the ground.

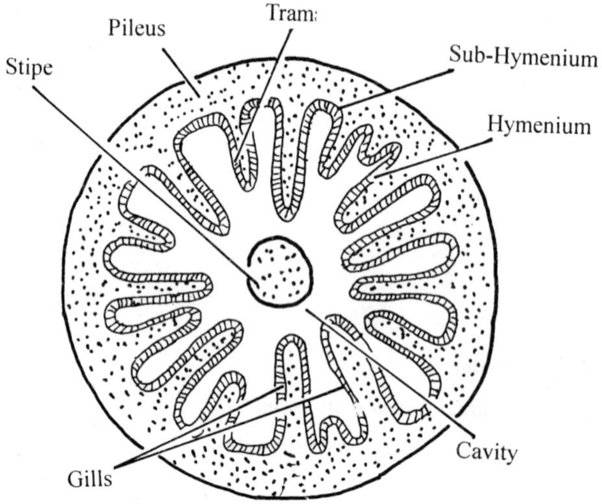

Fig. 3.7 Agaricus– Cross-section of Pileus

The basidiocarp :

1. The basidiocarp is formed by the underlying materials of the hyphae after dikaryotization.

2. The young basidiocarp, when it emerges out of the ground will be button shaped, hence it is known as the button stage.

3. The button appears as a small globose whitish structure above the ground.

4. In a long sectional view, the button shows a slightly expanded top portion separated from the lower ones or portion by means of a constriction.

5. A little above the contriction are found two chamber the lamellar cavities.

6. The margin of the pileus is attached to the stipe with the help of a thin membrane called veil or velum.

Mature basidiocarp :

1. The mature basidiocarp of Agaricus is a small umbrella structure.
2. It has a slong, dull white coloured, cylindrical stalk called the style.
3. The stipe bears at its tip an expanded portion called the stipe.
4. Midwary on the stipe is found the annulus representing the represents of the velum, which gets torn as the pilaus expands.
5. The upper surface of the pileus is smooth or undualted.

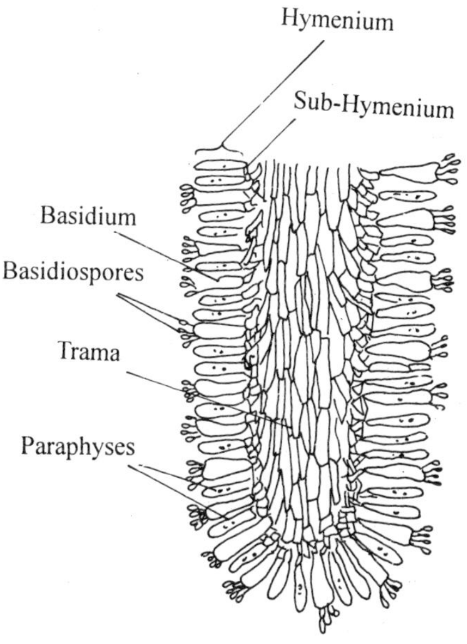

Fig. 3.7 Agaricus–V.S of a gill magnified

6. The undersurface of the pileus has got a number of plates or lamellae radiating from the centre. These lamellae are called gills.

7. The gills that extend right from the centre to the margin of pileus are called primary gills while those that start midway and reach the margin are called secondary gills.

Structure of a gill:

1. A vertical transverse section of the gill shows the following features :

2. At the central region are present loosely arranged fungal hyphae constituting the trama.

3. On either side of the trama are found two to three layers of compactly arranged hyphae which looks like rounded cells. This is called the subhymenium.

4. External to the sub-hymenium and arranged at right aneles to the flat surface of the till are found basidia constituting the hymenium.

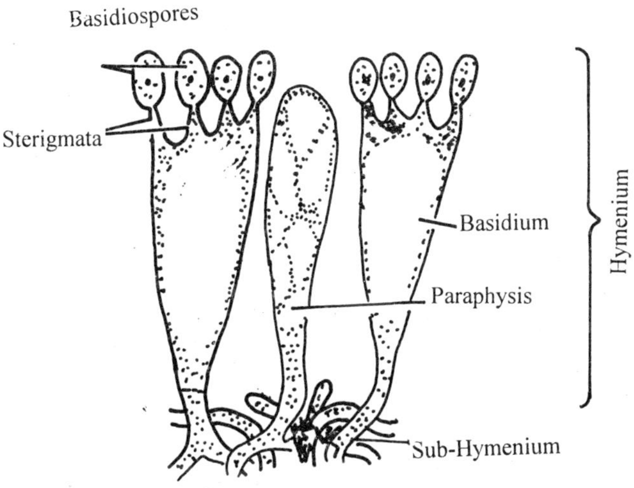

Fig. 3.8 Agaricus–Gills

5. The hymenium also has cystidia, the sterile hyphae intermingled with the basidia.

6. The basidia reproduction club shaped and aseptate. They are called holobasidia.

7. The basidiospores are small and oval in shape. They are wind sispersed and on the availability of a suitable substratum, they germinte to give rise to the monokaryotic primary mycelium of a new individual.

Specimens given in the laboratory :

1. Basidiocarp
2. Section of a gill (slide)

Diagrams to be drawn:

1. Basidiocarp
2. Section of agill
3. A few basidia enlarged.

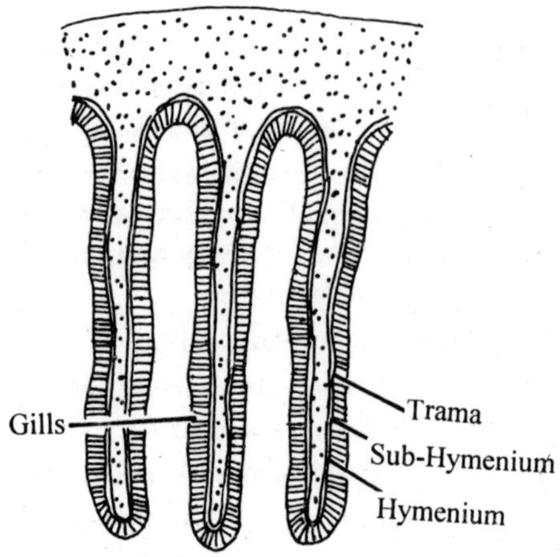

Fig. 3.9 Agaricus– Hymenium

CHAPTER 4

PLANT MORPHOLOGY II

STUDY OF POLYTRICHUM

Common species : *P.densifoulmin, P.commune, P.xanthopilum, P.juniperinum* etc.

Occurrence : The genus with about 100 species preference to cool temperate and tropical region. Cool shady soil, marches, peat bog, soil along pond banks etc are the usual habitats where polytrichum grows.

Study : 1 External morphology

2. Anatomy of stem
3. Anatomy of leaf
4. Sexual reproduction
5. Structure of the sporophyte
6. Protonema
7. Systematic position and identification

External morphology

1. The plant body of Polytrichum is a gametophyte.
2. It has an underground rhizome from which spring up leafy shoots.
3. In fair environmental conditions, the leafy may each a height of 20 cms.
4. The stem is erect and cylindrical. It is branched.
5. In P.commune, a traditional region is found between the rhizome and the arial shoots.
6. The leaves are spirally arranged in three vertical rows.
7. The leaves on the transitional zone and rhizome are brown or colourless, those on the aerial shoots are large and rigid.
8. The leaves are provided with a sheathing membranes leaf base which narrows up into a limb.
9. The leaf has a broad midrib, which bears on its adaxial face many cell plates (lamellae). These are one celled thick and photosynthetic.
10. The rhizome is anchored to the soil with the help of rhizoids.
11. The rhizoids are multicellular with oblique cross walls.

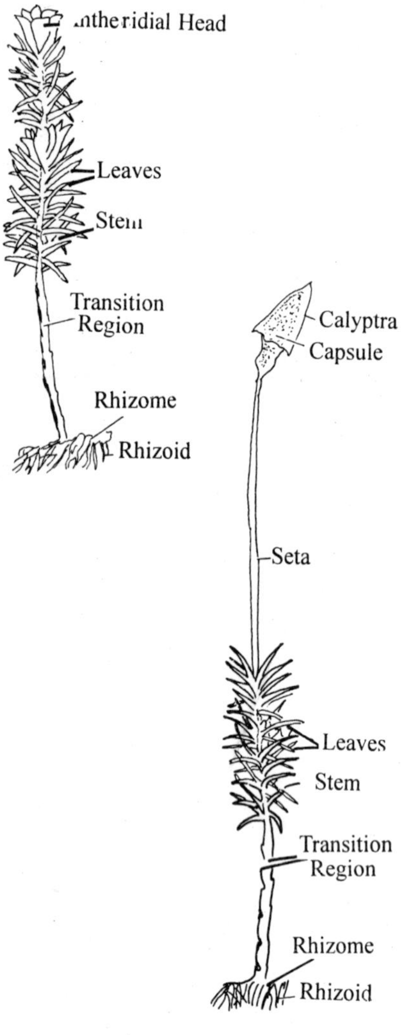

Fig. 4.1 Polytrichum–A female plant

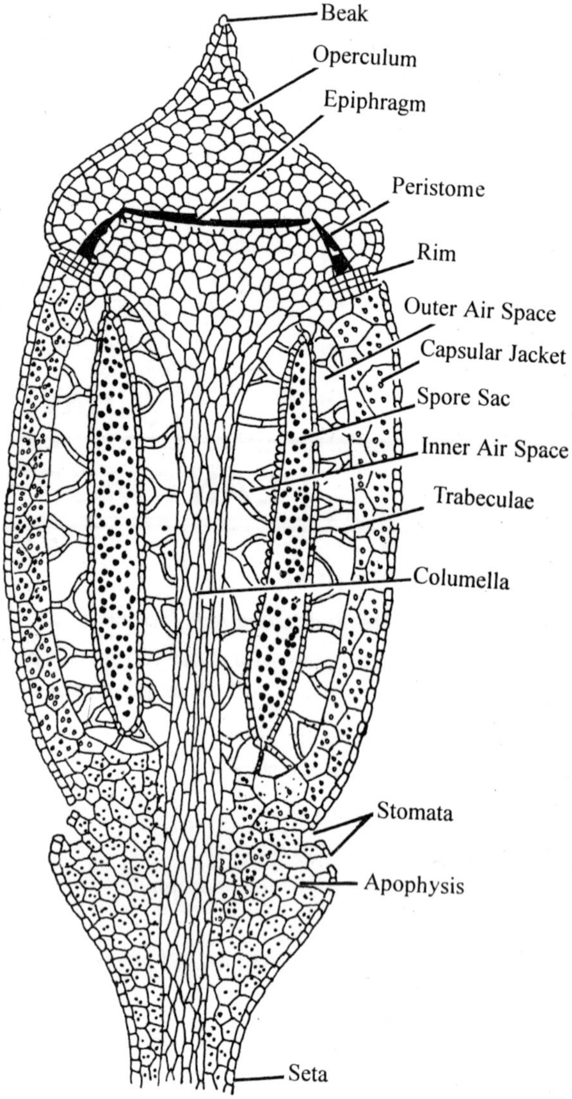

Fig. 4.2 Polytrichum–L.S. of capsule

12. The rhizoids entangle themselves into loosely knit manes, thus helping in the external capillary movement of water.

13. The water can rise through the wick like rhizoidal manes.

Sexual reproduction

1. Sexual reproduction is oogamous. Sex organs are antheridia (male) and archegoma (female).
2. Polytrichum is generally dioecious.

Antheridia

1. These are generally borne in terminal cluster forming open flower like cups.

2. The cup is surrounded by perigonial leaves that are very much different from the vegetative leaves.

3. In some cases, the vegetative shoot apex may grow through the antheridial cluster producing one more group of antheridia.

4. Within the cup are found a number of antheridia intermingled with paraphyses.

5. Each antheridium is a clubshaped structure having a short stalk.

6. The jacket of the antheridum is made up of a single layer of cells.

7. The antheridium has a number of androcytes which metamorphose to give rise to coiled biciliate antherozoids.

Archegonia

1. Archegonia are also borne on terminal clusters.
2. There are surrounded by pink or rare coloured perichaetical leaves.
3. The leaves are arranged in a rosette looking like a flower.
4. Usually three archegonia are found in each cluster as in P.commune.
5. With the development of the archegonia, the growth of the shoot stops.
6. Scattered among the archegonia are found a number of hair like structures called paraphyses.
7. Each archegonium has a stalk, a bulbous venter and a spirally twisted neck.
8. Rest of the structural details of archegonia are same as seen in Funaria.
9. Fertilization is as seen in Funaria.

Sporophyte

1. The sporophyte has three parts viz., basal foot, middle seta and the terminal capsule.
2. The foot is parenchymatous and is embedded in the archegomal cluster.
3. The seta is long and cylindrical. Anatomically it shows the following features.
4. The outermost layer of cells is thick walled.
5. Internal of this is band of sclerotic cells which gradually merge into the inner thin walled loosely arrangement chlorophyllous cells.
6. In the centre of the seta ruus a conducting strand.
7. The seta terminates at the base of the capsule into a swollen structure called the apophysis.
8. The aphophisis is separted from the rest of the capsule by a constriction.
9. Automatically the apophysis has an epidermis interrupted with stomata.
10. The epidermis encloses a mass of green chlorenchymatous cells.
11. In long section, the capsule shows an outer wall composed of several layers of cells.
12. The outermost layer of the wall is the epidermis.
13. Internal to the wall is an airspace (outer lacuna) separated into air chambers by means of parenchymatous partitions.
14. The cells of these partitions are chlorophyllous and are connected internally to the outer wall of the spore sac.
15. Next to the outer air chamber is found the spore sac which is covered on either side by means of a 2 layered wall.
16. The wall encloses the archesporium of 4-6 layers of cells which undergo reduction division to form haploid spores.

17. There are no elaters.

18. Internal to the spore sac is the inner air space (inner Lacuna) which is similar in structure to the outer air space.

19. The partition walls of the inner air space are attached to the central columella.

20. The columella consists of vertically elongated parenchyma cells.

21. The capsule is covered on top by a lid like operculum.

22. The operculum has a beak or rostrum.

23. The annulus is not distinct like in Funaria.

24. The operculum at its base has the epiphtragm. This is made up of a transverse band of thin membranes cells without intercellular spaces.

25. This stretches like a drum head (tympanum) covering the opening of the capsule.

26. Outside the epiphragm is the peristome.

27. The peristome has 36 - 64 teeth which are connected to the capsule below and to the epiphragm above.

28. The spores are minute in size, haploid and uninucleate.

29. On the availability of a suitable substratum, the spores germinate to develop into protonema.

Protonema

1. The spores germinates to form a branched, septate filamentous protonema.
2. Some of the branches grow vertically above (chloronema) while some others grow down and help in anchorage.
3. Subsequently buds appear near the base of chloronema.
4. These buds develop into leafy gametophytes of polytrichum.

Diagrams to be drawn

1. Habit sketch
2. Anatomy of rhizome
3. Anatomy of aerial shoot
4. Anatomy of leaf
5. L.S of male flower
6. L.S of female flower
7. L.S of sporophyte
8. Protonema

STUDY OF NEPHROLEPIS

External features

The sporophytic plant body is differentiated into stem, root and leaves. The stem is erect, condensed and is commonly called the rhizome or the root stalk. The rhizome is branched. The branches are long, slender and are called runners. At its lower face the rhizome bears the roots. The runners spread to a wider area and bear roots. The runners help in vegetative propagation, as they bear buds. These buds give rise to new leaf bearing root stalks. Usually there will be runner at the base of each leaf.

The rhizome as well as the runners are covered with peltate scales and branched hair like structures called ramenta.

The leaves of Nephrolepis are vere characteristic. They are commoly referred to as the frond. They are long (25-30 cm) slender and have a long rachis. The leaves when young show circinate vernation ie:, the young growing tip is coiled inwards and open up during further growth. A mature leaf is plnnately compound.

The pinnae are slightly falacate and have free veins. The veins at their tips have white lime dots. All parts of the leaves are covered with scales.

The pinnae on their lower surface bear aggregation of reproductive structures called sori (sing.sorus). The sori may be roundish or elongate and are present at the tips of the veins on the pinnae.

Anatomy of Rhizome

Internally the rhizome shows an epidermis, cortex and stele. The epidermis is single-layered, while the cortex is mostly parenchymatous.

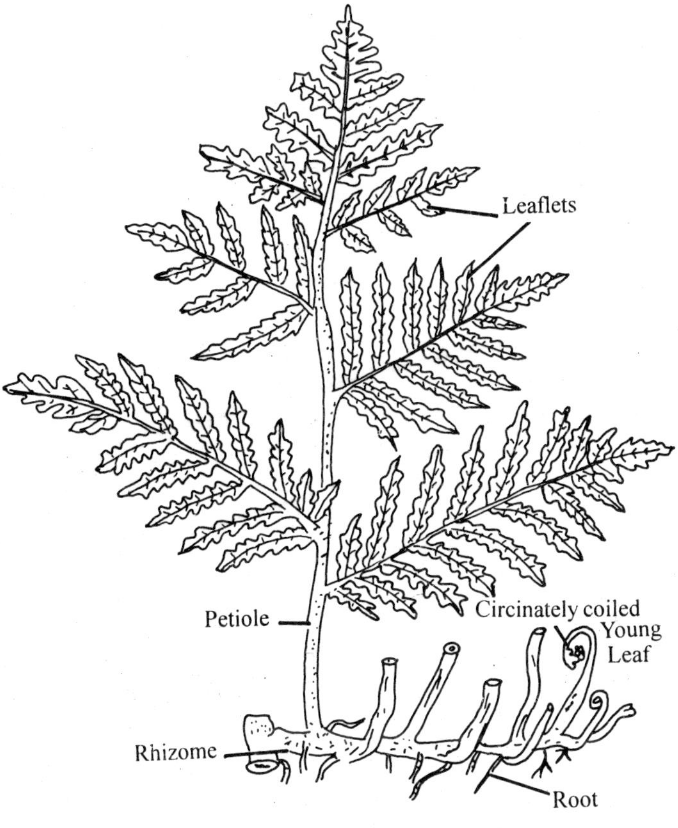

Fig. 4.3 Nephrolepis–Habit of plant body (External features)

The central vascular cylinder is said to be dictyostele. It consists of many independent units of vasculature called the meristeles.

Reproduction

The sporophyte reproduces by sporangia which are aggregated into sori. A cross section of the pinna passing through a sorus will reveal the details of the sporangium.

Structure of a sorus

The sorus at its basal region has a group of fertile tissue called the receptacle. The sporangia develop from the cells of the receptacle. Generally in a sours one can see a number of sporangia at different stages of growth. The entire sorus is covered by a flap of tissue called the inducium. The function of the induculum is to protect the developing sporangia. The inducium in Nephrolepis is usually round or reniform with a curved base.

Structure of a mature sporangium

The sporangium is somewhat a globose structure. It has multicellular stalk and a globose capsular region. The wall of the sporangium is made up of three types of cells.

1. Annular cells
2. Stomial cells
3. Parenchyma cells

The annular cells are thick walled and they incompletely overarch the sporangium. The stomatal cells are thin walled and radially elongated. They are located at the point where the annulus stops. They are at right angles to the stalk of the sporangium. The rest of the wall is made up of ordinary parenchyma cells. The sporangium consists of diploid spore mother cells which undergo reduction division to give rise to haploid spores. All the spores are of the same type (homospores).

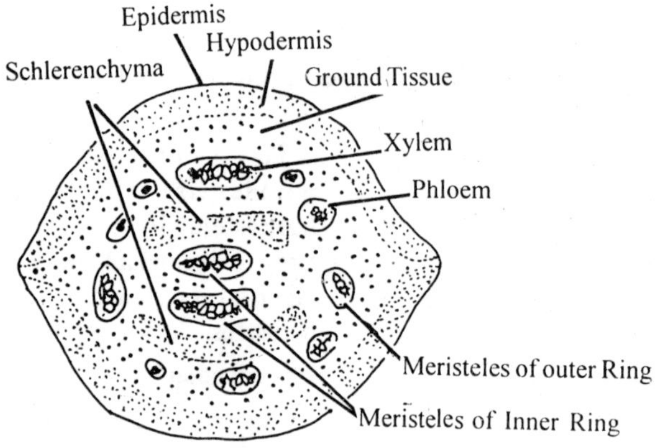

Fig. 4.4 Nephrolepis–T.S. of rhizome (diagrammatic)

Dehiscence of the sporangium

The annular cells are hygroscopic. As a response to the moisture content in the atmosphere they shrink. As a result of this, the stomatal cells are torn apart the sporangium breaks open transversely exposing the spores. The spores come out and are dispersed by wind.

The spores are minute and compressed laterally. On falling upon a suitable substratum the spores develop into a gametophyte or prothallus.

Structure of the prothallus

The prothallus or gametophytes of Nephrolepis is small (1/8" in diameter) and hardly visible to the naked eye. It grows on moist, cool and shaded soil generally rich in organic matter.

The gametophyte is thin and made up of parenchyma cells. It is cordate in shape. The cells are chlorophyllous. The prothallus is one cell thick towards the maring, but few cells thick towards the centre. The growing point of the gametophyte is situated in the apical notch. The ventral surface of the thallus has a number of rhizoids which help in anchorage and absorption.

The prothallus is monoecious. It has both antheridia (male) and archegonia (female) on the same plant body. The antheridia are found on the ventral surface, distributed among the rhizoids, while the archegonia are present near the apical notch facing the antheridia.

Structure of antheridium

The mature antheridium is globose and is embedded in the thallus. It has a single layered jacket (wall). Inside the antheridium are found 20-50 large, spirally coiled antherzoids. At maturity the wall breaks open releasing the antherizoids. The antheridium of ferns is a much simpler reduced structure when compared with that of bryophytes.

Structure of the archegonium

In the structure of the archegonium resembles that of Funaira, but it is a much reduced structure. It has a bulbous archegonium are found a long neck canal cell, a venter canal coil and an egg cell. In a mature archegonium all these cells dissolve except the egg cell.

The disintegrating cells of the archegonium ooze out through the neck. They probably contain a chemical substance which attracts the antherozoids. Finally only one succeeds in fusing with the egg resulting in the formation of zygote.

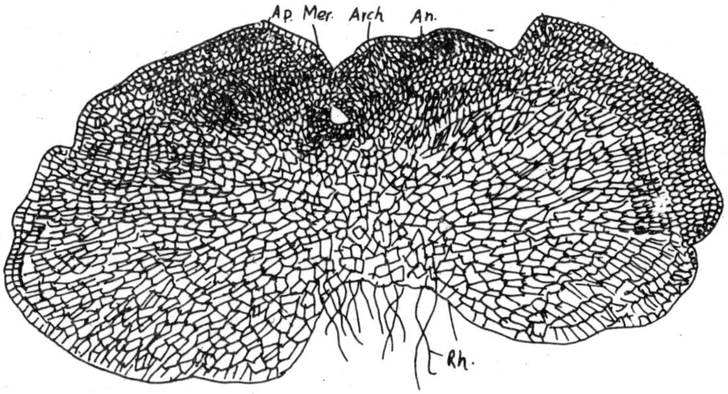

Fig. 4.5 Gametophyte of Nephrolepis

The zygote develops into sporophyte. From the zygote, first an embryo results. Later this embryo gives rise to the characteristic fern plant or the sporophyte.

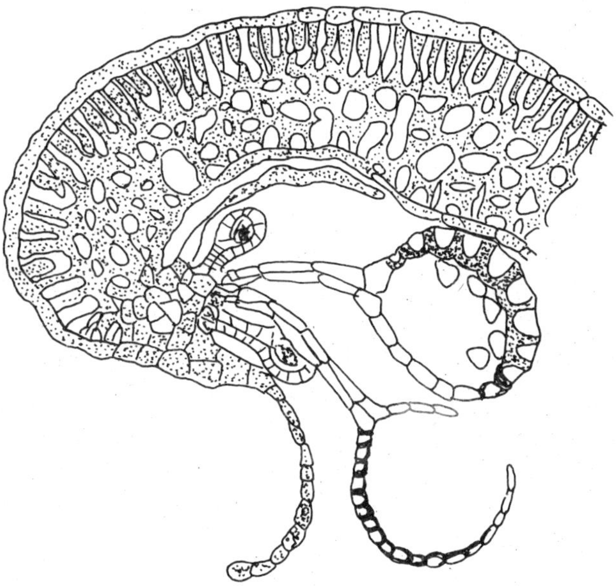

Fig. 4.6 Section of pinna through a sorus

STUDY OF CYCAS

Cycadales

Cycadaceae

Cycas

The plant body consists of an aerial, erect stem which is anchored to the soil with the help of tap root as stout as the main stem. The plant reaches a height of 3 to 6 metres resembling a small palm tree. Some of the species may reach a height of 20 meters. The stem at its tip bears a crown of graceful leaves. At the base, the stem has a diameter of 0.5 to 1 metre. It has a thick armour of persistance leaf bases. The stem is generally unbranched but branched specimens are not very rare. Branching is dichotomous when it occurs. Very rarely three branches may arise at a point.

The leaves are borne at the apex of the stem in a spiral manner. They are of two types, viz., scale leaves and foliage. In the axils of these, the foilage leaves develop. The foilage leaves are more than a metre long when fully grown and exhibit circinate vernation which is unusual to Gymnosperms. The leaves are pinnately compound and have a stout rachis. Some of the basal pinnae are modified into spines. The pinnae are linear and have a central unbranched vein.

The stem is anchored to the soil with the help of a tap root which is as stout as the stem. The tap root bears a number of secondary roots. Some of these secondary roots are modified into coralloid roots. These are irregularly branched cluster of roots that grow apogeotropiclly and are found a few inches beneath the soil. The coralloid roots are greyish green in color and owe their name to the irregular coral like branching. On the surface of the root, smalll lenticel like openings are found probably helping in respiration. Internally, the coralloid roots have an endophytic blue-green alga (Anabena cycadacearum) in the cortical region. The relationship between the alga and the roots of Cycas is believed to be symbiotic. While alga provides nitrogenous material (it has nitrogen-fixing capacity) to Cycas, it gets protection and shelter in turn.

Male reproductive structures

The male cone is a compact structure and measures in length from 40 to 50 cm. It has a central cone axis on which are found spirally arranged leaf like structures called microsporophyils. Each microsporophyil is a flattened wedge shaped structure consisting of a narrow lower portion which broadens above. The upper position is drawn into a projection called apophysis. The microsporophyils on their abaxial (lower) surface bear a number of pollen sacs or microsporangia. The microsporangia are present in two large groups separated by a median sterile line. In each group, there are a number of aggregations of sporangia called sori. Each sorus has an elevated central porition around which are found 3 to 6 microsporangia. Between the microsporangia inducial haris are found. In each microsporangium a large number of dusty portion is produced.

Female reproductive structures

In the female plant there is no compact cone or strobilus as in the male plant. At the apex of the stem, the megasporophylls are spirally arranged. They develop between foliage leaves and scale leaves.

Each megasporophyll has two parts viz., a basal stalk like portions and an upper broad leaf like portion with a number of dentations. The megasporophylls resemble the foliage leaves and are believed to be modification of the latter. On the lower stalk like portion of tmegasporophyll two rows of ovules are borne. The number of ovules per megasporophyll range from 2 to 10. The ovules and the sporophylls are covered by yellowish woolly outgrowth.

The ovule of Cycas is of the size of a hen's egg and is green or orange red in color. It is the biggest ovule in the plant kingdom.

Structure of a mature ovule

In a long sectional view the detailed structure of the ovule will be clearly seen. The ovule is surrounded by an envelope called the integument. The integument is very thick and is made up of three layers, viz., outer fleshy layer (outer sarcotesta), middle stony layer (sclerotesta) and inner fleshy layer (inner sarcotesta). At

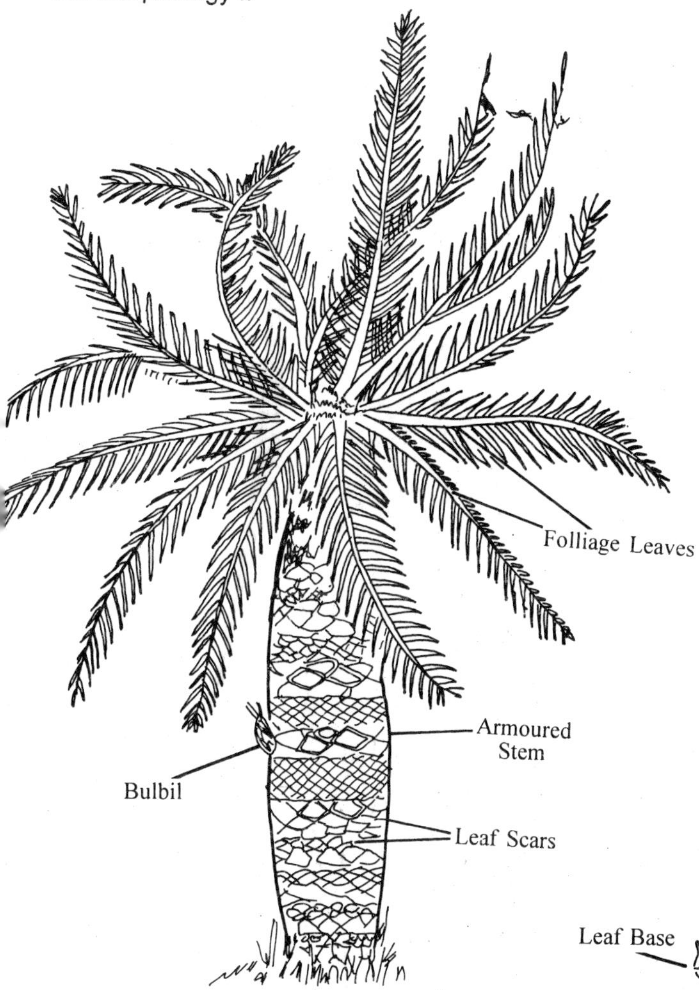

Fig. 4.7 Cycas–Habit of plant body External features

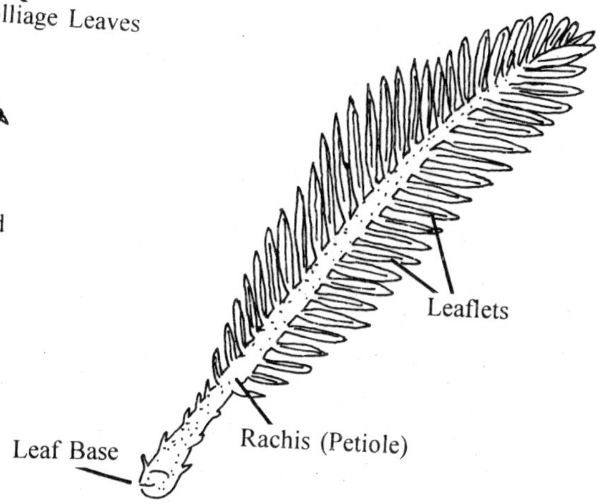

Fig. 4.8 Single leaf

one end (free end) the integument leaves a small opening called the micropyle. Since the integument is very thick, the micropyle forms a narrow tube called the micropylar canal. Internal to the integument is the nucellus (in a young ovule the inside of the ovule has only nucellus). Later the female gametophyte or embryo sac, develops from it. Internal to the nucellus, is the large cellular gametophyte occupying most of the interior of the ovule. The cells of the embryo sac are rich in starch content. The embryo sac is very often called the endosperm since it provides nutrition to the embryo (but this is different from the endosperm of angiosperms). The embryo sac is adhering to the nucellus at all regions except at the micropylar region where it leaves a small space called the arechegonial chamber. In the micropylar region of the embryo sac the archegonia are seen. Each archegonium has 2 neck cells, one ventral canal nucleus and one egg cell. The nucellus protrudes into the micropylar canal and forms the nucellar beak. Some cells of the nucellar beak disingetrate and form the pollen chamber is which the pollen grains are lodged during pollination.

STUDY OF THE PARTS OF A TYPICAL FLOWERING PLANT

Dicot plant (Mustard)

1. The plant body has two parts- the aerial shoot system and the underground root system.

2. The root system is a tap root system. There is a main primary root which gives rise to secondary and tertiary roots.

3. The shoot system comprises of stem, leaves, flowers and fruits.

4. The stem is cylindrical. It is divided into nodes and internodes.

5. At the apex the stem has a terminal bud.

6. Leaves are bound at the nodal regions. They have a petiole, lamina and leaf apex. Venation is reticulate.

7. Axillary buds are seen at the nodes in the axils of the leaves.

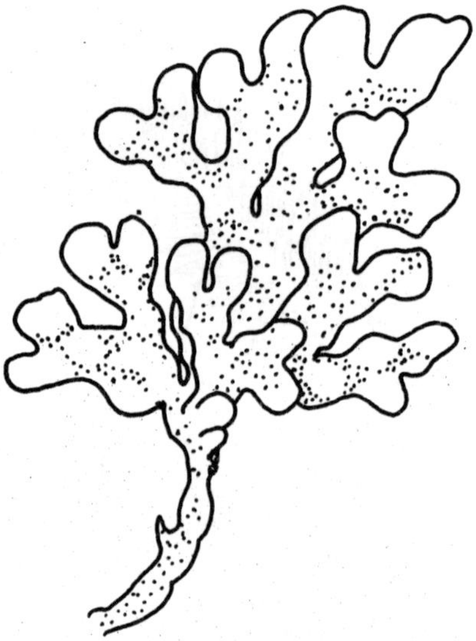

Fig. 4.9 Coralloid root

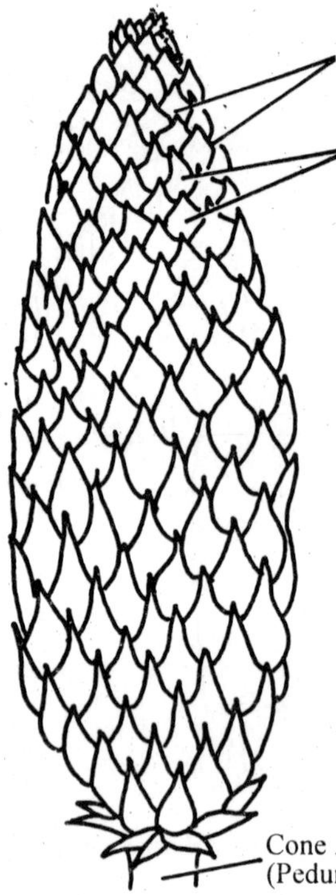

Fig. 4.10 Cycas–Male cone (External features)

8. Flowers are yellow in colour and are arranged in raceme.

9. Fruits are seen at the base of the inflorescence.

Monocot plant (Grass)

1. Same as in dicot.

2. Root system is adventitious, it is also called fibrous.

3. The shoot system has leaves, flowers and fruits.

4. Stem is divided into nodes and internodes, but it s covered by sheathing leaf base.

5. Leaves have a prominent sheathing leaf base, and leaf apex. There is no petiole. At the junction of leaf base and lamina there is a membraneous structure called ligule. Venation is parallel.

6. Flowers are arranged in spikelets.

CHAPTER 5

ROOT MODIFICATIONS

ROOT

The primary axis of a flowering plant is usually divided into two parts namely root and shoot. Root is the underground part while shoot is the aerial part. Occasionally however these may be aerial roots (Ex. orchid) and underground stems (Ex. potato).

Root usually grows downward into the soil and is achlorophyllous. It is positively geotropic and positively hydrotropic (growing towards moisture) also.

Types of Roots

Usually two types of roots systems are identified based on origin, type of growth etc. These are the Tap Root System or Primary Root System and Fibrous Root System or Adventitions Root System. The former is found in Dicotyledons and the latter in Monocotyledons.

The differences between the two types of Root Systems are as follows:-

	Tap Root system	Adventitious Root System
1.	Found in Dicots	Found in Monocots Radicle is short lived, Extra down and arise and constitute the fibrous roots.
2.	The radicle of the embryo grows forms the primary root. This remains for the entire life of the plant.	
3.	The root system has a prominent primary root on which arise secondary and tertiary roots.	No such distinction is seen. All roots seem to arise from a single point.

Normal functions of the Root

The root has two main functions. These can be categorized as mechanical functions and physiological functions. Mechanical function is the fixation of the plant firmly to the substratum. Physiological functions are absorption of water and mineral nutrients from the soil and transporting them up to the aerial parts.

Modification of Roots

Under normal conditions the roots perform the normal functions of fixation and absorption but when subjected to special environmental conditions the roots take up various other functions. In order to perform these varied functions the structure of the root also gets altered. All these alterations come under the purview of Root Modifications.

Broadly speaking, the root modifications fall under two categories viz., roots modified to perform special physiological functions and roots modified to perform special mechanical functions.

Roots Modified to perform physiological functions

The modified roots perform the following special physiological functions viz. Food storage, parasitic absorption of nutrition, epiphytism, respiration, and mycorrhizal functions.

1. **Food storage:** In many instances the food prepared by the aerial parts are stored in the roots. Both tap root and adventitious roots get modified to store the food.

Tap Root: Modified tap roots store the reserve food. Tap root modifications are classified into the following three types.

Fusiform: In this, the root is spindle like with both ends tapering and the central region swollen. The storage tissue is usually the parenchyma of the secondary xylem. Ex.*Raphanus sativus* (Raddish)

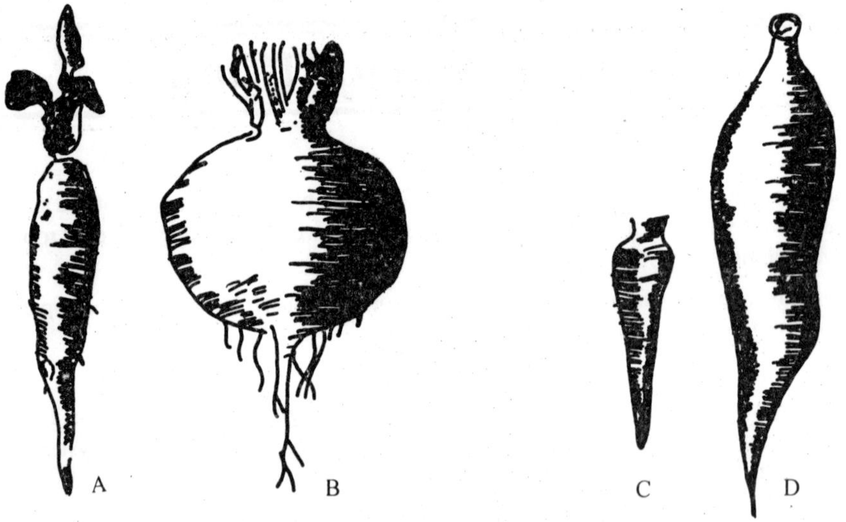

Fig. 5.1 Root - Tap root modifications
A. Fusiform (Raddish) **B.** Napiform (Beet root)
C. Concial (Carrot) **D.** Tuberous (Mirabillis)

Concial : In this type, the roots are conical with a broad top and a pointed base. In this plant food is stored in the cortex and parenchyma of secondary phloem. Ex. *Daucus, carrota.*

Napiform : Here one end is very broad and swollen and it suddenly tapers at the other end, Ex. *Beta vulgaris, Brassica rapa* (Beet root and turnip).

Adventitious roots : Adventitious roots also get modified to store the food material. They are classified into the following types based on their shape and structure.

Fasciculated : At the base of the stem a number of adventitious roots arise. All of them are swollen due to the accumulation of food material. The food is stored in the parenchyma. As the swollen roots are present in a cluster they are said to be fasciculated roots. Ex. *Dahlia, Orchids etc.*

Annular: The roots appear like aggregation of rings or annuli. Ex. *Ipecac, Orchids etc.*

Moniliform: The roots appear like a chain of beeds, with the food storing part being swollen. Ex: *Momordica charantia* (Bitter gourd)

Fig. 5.2 Root - Adventitious root modification
A. Tuberous (Sweet potato) **B.** Fasciculated (Dahila) **C.** Nodulose (Mango Ginger)

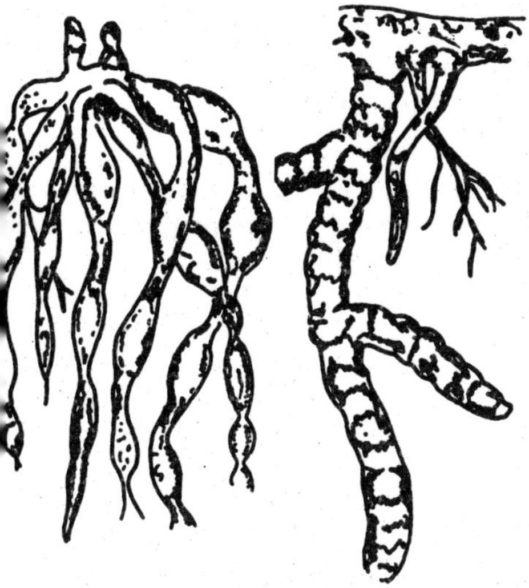

Fig. 5.3 A. Root - Adventitious root modifications A. Moniliform (Bitter gourd) **B.** Annulated (Ipecac)

Nodulated : Here the root tips get swollen into small nodules due to the accumulation of food. Ex: *Ginger, Melilotus* etc.

Tuberous: This type of storage adventitious root is found in *Ipomea batatus* (sweet potato). In this plant which is a creeper the adventitious roots that arise at the nodes accumulate food material.

2. **Parasitism :** Some angiosperms like *Cuscuta* or *Cassytha* are total parasites. They are climbing plants which twine round the host plant with the help of their yellowish stem. They lack leaves. The stem here and there produces certain special roots called *parasitic roots* or *haustorial roots.* These roots penetrate the host stem and reach the vascular bundles from where they absorb the food.

3. **Epiphytism:** Certain plants like *orchids* grow on the crevices of branch of tall trees. These plants are not parasites, even though these grow on other plants. Plants of this type are called *Epiphytes (epi=above, phyta=plant).* In such plants special roots called *hanging roots* or *epiphytic roots* are produced from the base of the plant. These roots hang down in the air and absorb moisture from the atmosphere. In order to help absorb moisture, epiphytic roots have special tissue called *velamen.* This spongy tissue absorbs water. The cortex of the epiphytic roots are green due to the presence of chloroplasts. They also help the plant in photosynthesis. As such they are also called *photosynthetic roots* or *Assimilatory roots* (they assimilate CO_2).

Fig. 5.4 Root-Epiphytic roots of *Vanda*

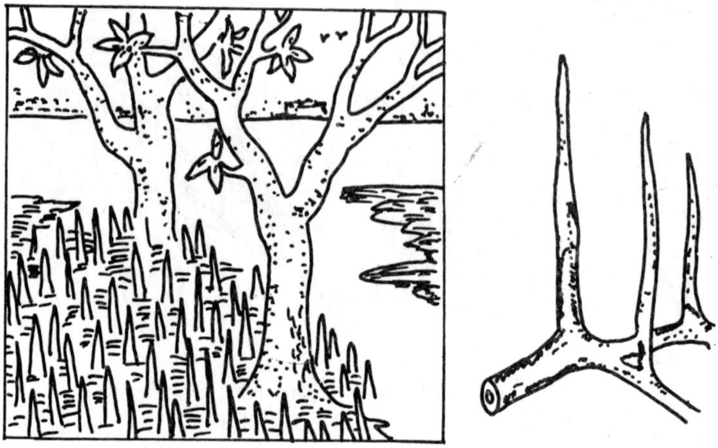

Fig. 5.5 Root- Pneumatophores of Avicennia
A. Plant with pneumatophores **B.** A root enlarged

4. Respiration: Many of the flowering plants grow in the backwaters of the sea. Here the plants are rooted firmly but the soil is submerged in salt water. Such plants are called *mangroove plants* (Ex: *Avicennia, Rhizophora, Salicornia etc).* In these plants the soil is poorly aerated as it is filled with water and the roots will not get enough oxygen for their respiration. Hence some of the roots produce special branches called *Pneumatophores* or *respiratory roots* or *breathing roots.* These roots grow apogeotropically (grow against gravity) come above the water and absorb oxygen from the atmosphere. In order to facilitate the absorption of O_2, pneumatophores have small pores called *air pores* or *breathing pores.*

5. Mycorrhiza: In certain plants like *Monotropa, Ranunculus acris* etc, the roots have an association with funcal hyphae. Such roots lack hairs and the hyphae are supposed to help in absorption. The association is said to be symbiotic (Fungi get food from roots).

6. Reproduction: In certain plants like sweet potato the tuberous roots arising at the nodes have adventitious buds. On separation, each bud is capable of growing into a new plant.

Roots modified to perform special mechanical functions

In these modifications, the roots perform varied mechanical functions like fixation, climbing, floating supporting etc. The following are some of the modifications.

7. Prop Roots: In certain trees like *Banyan* where the branches are very long and grow for a long distance from the main stem, the main stem will not be able to support the branches. Hence the branches develop special roots which grow down, reach the soil and offer support like pillars. These are called prop roots.

8. Climbing Roots: Certain plants like *Pothos (money plant), Piper* (Betel leaf) etc, have very weak stems; and cannot grow erect on their own. Hence these plants take the support of other erect growing plants. In order to get a firm hold on the supporting plant, the climbers produce special roots at their nodal regions. These climbing roots clasp the support and help the plant to climb.

9. Stilt Roots: Plants like *Pandanus* and *Maize* grow in loose soil. As such the usual root system will not be enough to allow the stem to stand erect. Hence the stem, (a little above the ground) produces special root which grow obliquely, reach the soil and support the plant.

Fig. 5.6 Root- Modifications for mechanical functions
A. Prop. roots of Banyan **B.** Stilt roots of Pandanus

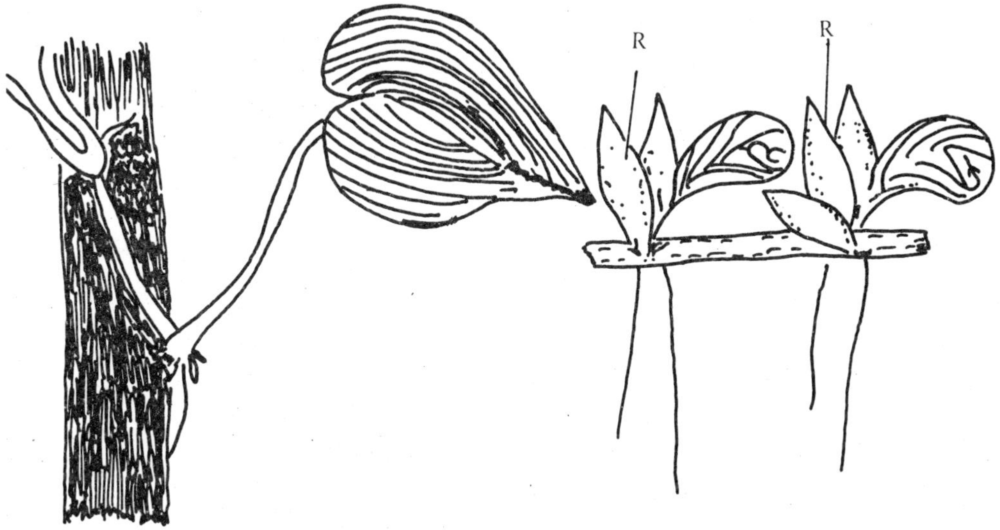

Fig. 5.7 Roots-Modifications for mechanical functions. **A.** Climbing roots of Piper (betel).
B. Floating roots of Jussieua.

10. Buttress Roots: In certain trees like *Ficus elastica, Eriodendron, Bombax* etc, the stem at its base produces flat wing like roots which radiate out and offer support to the trunk. These perform the same function as stilt roots.

11. Ballasting Roots: Water plants like *Eichhornia, Pistia* etc. are very light due to the presence of plenty of aerenchyma tissue. This is necessary for buoyancy. But at the same time this poses a problem also. If the plants do not have enough weight, they are likely to be blown away by wind. Hence the stem produces clusters of hair like roots which help to keep the plant in a position by offering adequate weight.

12. Contractile Roots: These roots develop in plants with underground stems like bulbs, corms etc. In order to keep the stem at a particular level in the soil, the roots grow out (from the underground stem). The contractile roots pull down the plant and keep it at a specific level in the soil. Ex. *Typhonium, Scilla, Crocus* etc.

13. Floating Roots: *Jussieua* is a hydrophyte floating on the surface of water. In order to help in buoyancy, the nodal regions of the plant produce short, white colored roots which store plenty of air. These roots help the plant in floating and also in respiration as they have air stored in the aerenchyma tissue.

CHAPTER 6

STEM MODIFICATIONS

STEM

In Angiosperms the primary axis of the embryo has two parts viz. the shoot and the root. The shoot is generally aerial while the root is underground.

The aerial part of the primary axis is called the shoot and it has two parts viz. the stem and the appendages. The axis together with its branches is called the stem, while the leaves, scales etc, constitute the appendages.

The stem is generally cylindrical, erect, positively phototropic and negatively geotropic. Stem is positively phototrophic for proper exposure of leaves to sunlight and to properly channelise the movement of fluids from and to the leaves.

Here and there on the stem are found certain swollen regions. These are called the nodes. The region of the stem in between the nodes is called *internode*. At the nodal regions usually the stem bears the leaves and the buds. The buds are present in the angle (axil) formed between the petiole and the stem. These buds are called *axillary buds* and are of two types (based on their function) viz. vegetative buds and reproductive buds. Vegetative buds develop into branches while the reproductive buds develop into flowers.

The apex of the stem terminates in a terminal bud. The terminal bud remains active usually throughout the life of the plant and contirbutes to the linear growth of the stem.

Normal functions of the stem.

1. It bears the appendages viz. leaves, buds etc.
2. It helps in the two way transportation of nutrients.
3. By being positively phototrophic, the stem helps in the proper exposure of the leaf surface to sunlight.

Modifications of stems

Under certain special circumstances stem may have to perform functions other than the normally assigned ones. In such situations depending on the modified functions, the structure of the stem also gets modified. Such modified structures are called stem modifications.

Stem modifications are usually of the following three types, which is mainly based on the location of the stem with reference to the soil. The three types of stem modifications are -

1. Underground (completely below the soil)
2. Sub aerial (partly below the soil)

3. Aerial (completely above the soil)

1. Underground Modifications

Underground stem modifications are of four types viz, Tuber, Rhizome, Corm, and Bulb.

Tuber : (Ex: *Solanum tuberosum* - Potato): In this modification, the stem or its branches grow into the soil. The tips become swollen due to the accumulation of food material. On the tuberous stem, buds may be seen. In potato the so called 'eye' are axillary buds, subtended by scale leaves. Each bud on sowing is capable of developing into an individual plant.

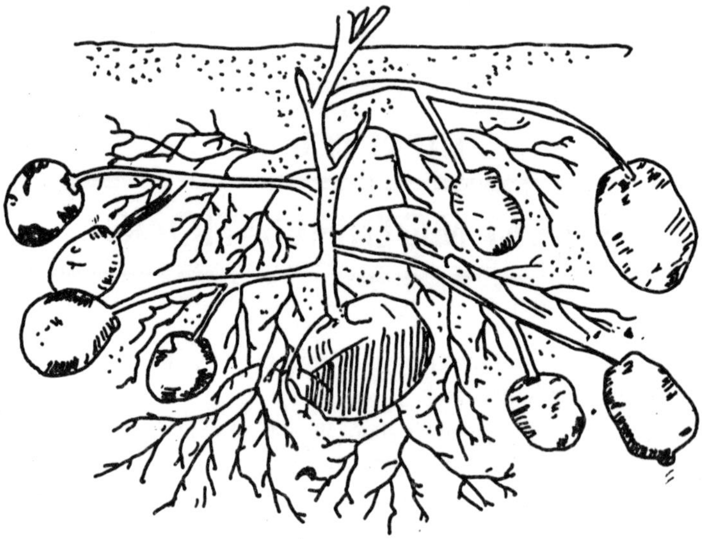

Fig. 6.1 Stem – Tubers of potato

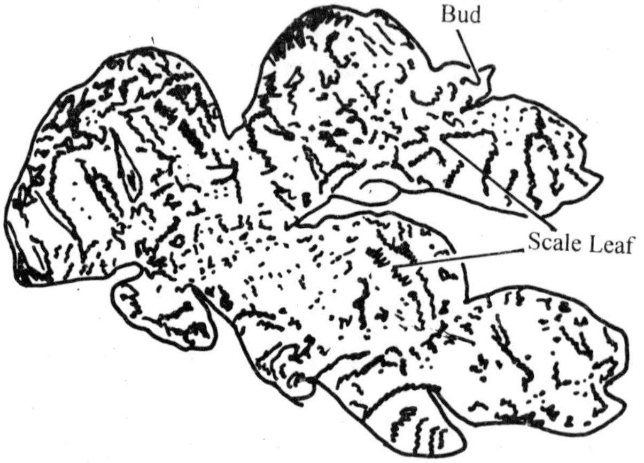

Fig. 6.2 Stem – Rhizome of ginger

Rhizome : In this type of modification the entire stem is underground and grows horizontally. The stem is divided into nodes and internodes with axillary buds and scale leaves at the nodes. The rhizome which is branched grows with the help of terminal bud. Food material is stored in the rhizome. During favourable season the buds develop into aerial shoots. The food prepared by the aerila shoots is stored in the underground rhizome. The aerial shoots die at the end of the favourable season while the rhizome persists in the soil. Ex: *Zingeber officinalis*- Ginger, *Carex* etc.

Corm : Corm is a globular solid stem that is completely underground. It is rich in reserve food material The whole stem is condensed. From the top of the corm, aerial shoots are produced, while the roots are produced from the base. The food prepared by the aerial shoots is stored in daughter corms produced by the axillary buds present on the main corm. Ex: *Amorphophallus* (Yam), *Colchicum, Gladiolus* etc.

Fig. 6.3 Stem – Corm of Gladiolus (A) and Amorphophallus (B)

Bulb : In bulbs, the actual stem is a highly condensed disc like portion, covered by fleshy storage leaves (Ex. Onion). The condensed stem gives rise to roots below. The bulb when sown gives rise to green, cylindrical foliage leaves which photosynthesise. The food prepared by the foliage leaves will be stored in the storage leaves. The bulbs are of two types based on the arrangement of the storage leaves on the stem. In *Concentric bulbs* (onion) the leaves are arranged in concentric rings without overlapping, where as in *tunciated bulbs* (Garlic) they overlap.Ex: *Allium cepa* (Onion), *Allium sativum* (garlic), *Tulipa Narcissus* etc.

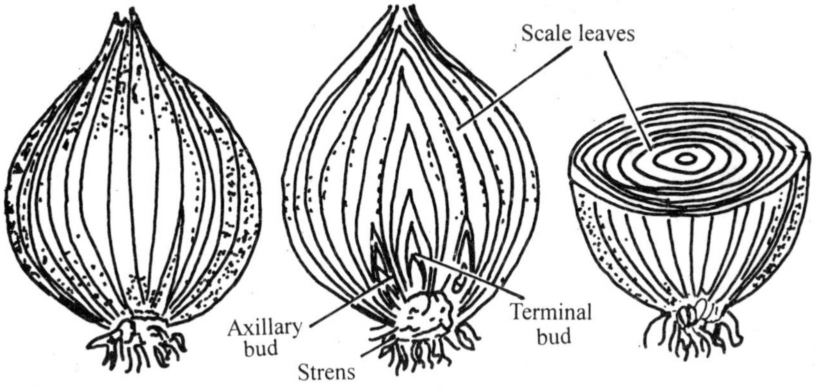

Fig. 6.4 Stem – Bulb of onion **A.** Entire bulb **B.** Long section of the bulb **C.** Cross section of the bulb

2. Sub aerial modifications

These modifications are partly underground, and generally grow horizontally. The sub aerial modifications develop from an axillary bud, which grows into a short branch. The tip of the branch produces roots below and leaves above developing into a new plant. The main purpose of sub aerial modification is vegetative propagation.

Sub aerial modifications are of the following four types-

Runner Ex: *Oxalis,* in this, one of the lowermost axillary buds on the stem develops into a short branch (runner).

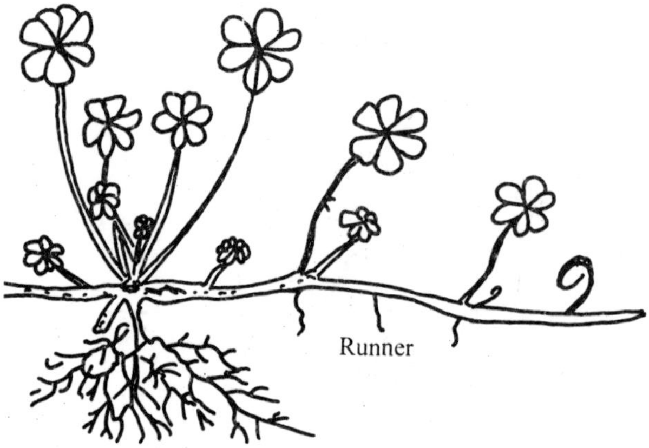

Fig. 6.5 Stem – Runner of *Oxalis*

This grows horizontally for some distance and then the tip gives rise to adventitious roots below and leaves above. This part may again give rise tc one more runner. In this way, the plant spreads rapidly. If for some reason, the runner decays, each part will become a new plant.

Stolon : Ex: *Colocasia*. A stolon is essentially similar to a runner except that it is shorter and mostly underground. Many botanists regard runners and stolons as one and the same.

Sucker : Ex: *Chrysanthemum, Mentha* etc.

Suckers are short branches that arise from the underground part of the stem. These grow up obliquely and come above the ground producing an aerial shoot. Sucker produces adventitious roots at the nodal region just below the soil. On being separated from the main plant a sucker can develop into a new individual.

Offset : Ex: *Eichhornia, Pistia* etc. Stolons produced in hydrophytes (water plants) are called offsets. An offset develops from the lower part of the plant and grows horizontally. The tip produces roots below and leaves above. On the decay or separation from the main plant, an offset grows into a new plant.

Fig. 6.6 Stem – Stolon of Fragaria

3. **Aerial modification**

These modifications are completely above the ground and have different purposes. The following are the main functions attributed to modified aerial stems.

Climbing: Certain portions of the stem are modified, in weak or slender stemmed plants so as to help them in climbing. Climbing modifications are of two types: Tendrils and hooks.

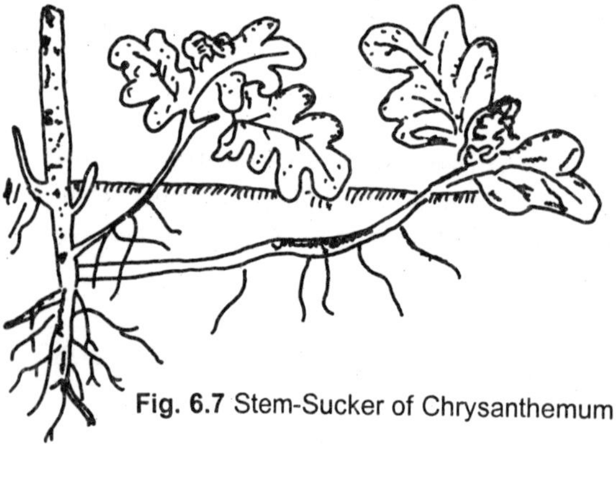

Fig. 6.7 Stem-Sucker of Chrysanthemum

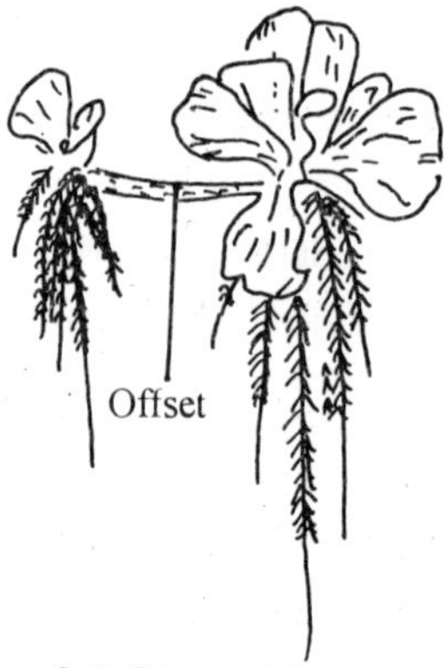

Fig. 6.8 Stem-Offset of Pisitia

Tendrils : Tendrils are long coiled, slender appendages that twine round the object which they come in contact with. Thus they help the climber to grow erect. Tendrils may be modifications of axillary buds as in *Passiflora edulis* or of terminal buds as in *Vitis quadrangnlaris*. In *V. quandrangularis,* the terminal bud gets modified into a tendril and further growth is brought about by an axillary bud. This grows up to one internodal lenght and the tip (bud) once again by the activity of an axillary bud. This results in zig-zag growth of the stem.

In *Antigonon,* a climber, the terminal part of the peduncle (inflorescence axis) and some of the terminal flowers themselves get transformed into a tendril.

Hooks : These are hard curved structures that attach themselves to a support, thus enabling the plant to climb. Hooks are found in *Artabotrys,* where it is a modified peduncle. In *Artabotrys,* while the peduncle gets modified into a hook, its tip bears a single flower.

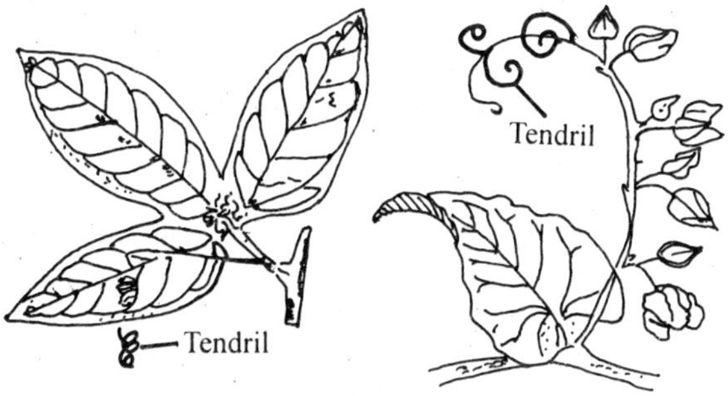

Fig. 6.9 Stem- Tendrils of Passiflora (A) and Antigonon (B)

Protection : In certain plants with tender and fleshy foliage, grazing animals prove to be a great hazard. Such plants develop sharp pointed structures, which discourage the animals from grazing on them. Such hard, pointed structures, which represent a modified stem are called *thorns* (Spines and prickles are also pointed structures, but different from throns).

Either an axillary bud *(Canthium)* or a terminal bud *(Carissa Caranda)* may become a thorn. In some cases, the thorn may be branched *(Flacourtia)* or may bear flowers *(Prunus)* at its tip.

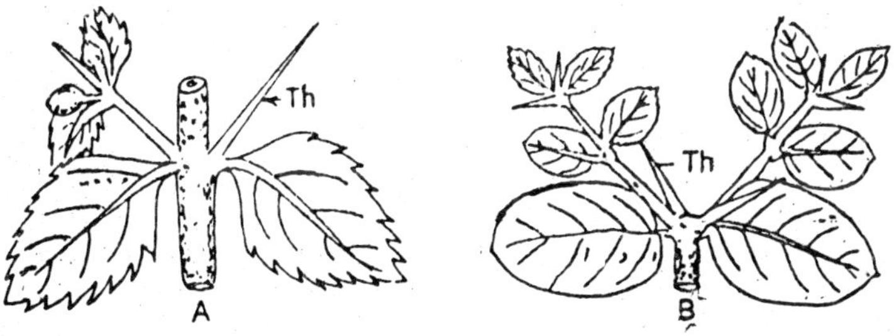

Fig. 6.10 Stem – Thorns of Durantha (A) and Carissa (B)

CHAPTER 7

LEAF

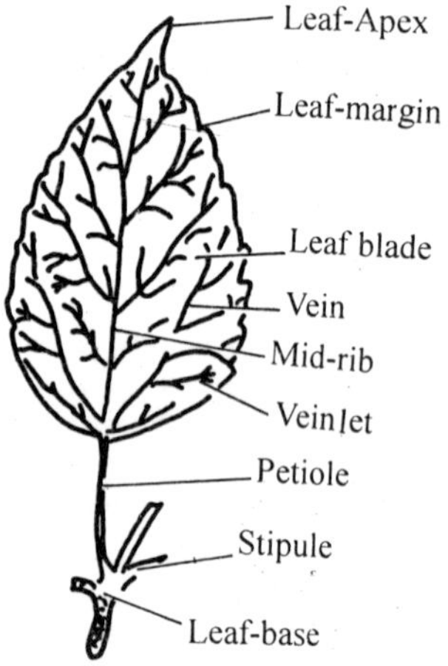

Fig. 7.1 Leaf – Parts of a typical leaf (Diagrammatic) (after Dutta)

LEAF

One of the most characteristic features of Angiosperms is their photosynthetic appendage-the leaf. The leaves are of various size, shapes and colors also.

A leaf is generally regarded as an outgrowth or appendage of the stem. An accurate definition of leaf, however would be that it is a dorsiventrally flattened photosynthetic appendage of the plant body bearing a bud in its axil.

Parts of leaf : Basically, a leaf has the following parts-(1) leaf base, (2) Petiole, (3) Lamina and (4) The leaf apex. Leaf base is the basal region of the leaf. Petiole is the stalk of the leaf. It helps to carry the lamina away from the stem. Lamina is the main part of the leaf. It is usually dorsiventrally compressed, and is also called the leaf blade. The tip of the leaf is called the leaf apex.

Differences between Monocot and Dicot leaf

	Dicot	Monocot
1.	Dorsiventral	Isobilateral
2.	Leaf base not prominent	Leaf base is prominent and forms a sheath surrounding the internode of the stem.
3.	Petiole generally present	Petiole generally absent
4.	Veination reticulate (network).	Venation parallel

Types of leaves

Leaves are classified into two basic types viz, **simple** and **compound**.

A simple leaf is one which consists of a single leaf blade. The leaf blade may be divided, undivided, lobed or dissected, if divided, the dissections of the lamina are not very deep, so as to separate it (lamina) into individual parts.

A compound leaf has several leaf blades. Each separate leaf blade may have its own petiole. Each one is called a leaflet (pinna). The stalk on which the leaflets arise is generally called the rachis. Sometimes, the rachis is branched in which case, the branch is called the secondary rachis. The main rachis is called the primary rachis. The individual leaflets though look like a leaf, do not have an axillary bud.

Fig. 7.2 Pinnately compound leaves. **A.** Unipinnate, (Paripinnate) **B.** Unipinnate (Imparipinnate) **C.** Bipinnate, **D.** Tripinnate

Types of compound leaf

A compound leaf may be classified under two categories. These are:-

1. Pinnately compound leaves and
2. Palmately compound leaves.

In pinnately compound leaves, the leaflets arise all along on the rachis.

The pinnately compound leaves are divided into 4 categories, based on the number of times the rachis is branched. These are :-

1. **Unipinnate :**The primary rachis is unbranched. Leaflets arise on the primary rachis. Eg. *Cassia.*

2. **Bipinnate:** The primary rachis is branched once. Leaflets arise on the secondary rachis. Eg. *Mimosa pudica.*

3. **Tripinnate:** The primary rachis is branched twice. Leaflets arise on the teritary rachis. Eg. *Moringa.*

Decompound : Here the branching of the primary rachis is more than three times or irregular. Eg. *Coriander.* Based on the number of leaflets in a rachis or its branch, a pinnately compound leaf is also calssified into two types. These are:-

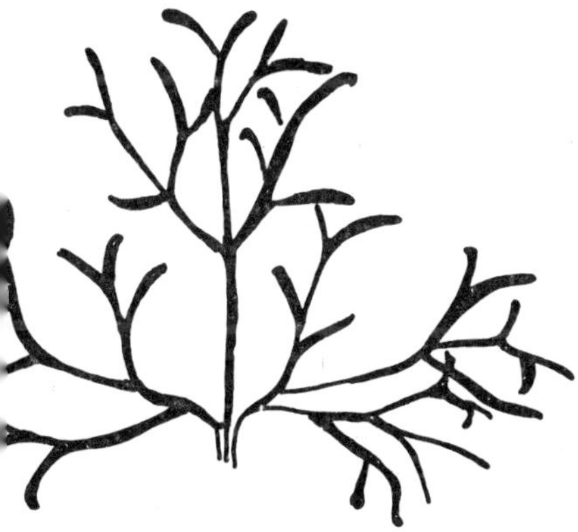

Fig. 7.3 Decompound leaf of Coriandrum

Fig. 7.4 Biloliately Compound leaf (Palmate)

1. ***Paripinnate :*** Even number of leaflets (rachis ends with a pair of leaflets).

2. ***Imparipinnate :*** Odd number of leaflets (rachis ends with a single leaflet)

The basis of classification in a plamately compound leaf is the number of leaflets. The number of leaflets may vary from one to nine or sometimes 12. The different types of palmately compound leaves are:-

1. ***Unifoliote:*** Attached to the rachis is a single leaflet. It appears like a single leaf, but the joint at the base is an indication of a compound leaf. Eg. *Citrus*.

2. ***Bifoliate*** : Number of leaflets two. Eg. *Hardwikia, Bauhinia, Balanites.*

3. ***Trifoliate*** : Leaflets three in number, Eg. *Crotalaria, Oxalis.*
4. ***Quadrifoliate*** : Leaflets four in number. Eg. *Marsilea.*
5. ***Digitate** or **Multifoliate*** :Number of leaflets five or more. Eg. *Ipomea spp.,Bombax etc.*

Fig. 7.5 Palmately compound leaves. **A-B.** Digitately compound leaves of Gynandropsis and Bombax **C.** Unifoliate leaf of citrus.

Normal functions of Leaf

1. It is the chief photosynthetic organ.
2. Much of transpiration is through the leaves.
3. Stomata present in the leaf help in gaseous Exchange during photosynthesis and respiration.
4. In monocots, leaf protects the growing apices.
5. Leaf also protects the buds in the young condition.

Modifications of leaf

Under special circumstances, leaf may have to perform various functions other than the normal ones. In such situations with the changed function the structure also will change and such changes are called leaf modifications. Leaf modifications are for the following purposes.

1. Climbing
2. Protection and xerophytic adaptation
3. Reproduction
4. Heterotrophic nutrition

Fig. 7.6 Leaf modifications. **A.** Tendrils in *Pisum sativum*. **B.** Tendrils in *Lathyrus*. **C.** Tendril of leaf tip in *Gloriosa*.

1. **Climbing :** In order to help in climbing, the different parts of the leaf get modified into a tendril which twines round the supporting plant.

In *Clematis*, the petiole is modified into a tendril while in *Gloriosa superba* the apex is modified into a tendril. In *Smilax* the two stipules at the base of the leaf become tendrils. In *Pisum sativum*, the leaf is unipinnately compound. Here the two stipules at the base of the leaf become leaf like (foliaceous) while the terminal leaflets become tendrils. In *Bignonia*, the central leaflet in the compound leaf is modified into a hook.

2. **Protection and xerophytic adaptation :** In xerophytic plants leaves are variously reduced and modified to suit the habitat. The following are some of the modifications.

In *Asparagus* the leaves are reduced to scales, while the stem takes up the function of photosynthesis.

In *Parkinsonia* of the family Caesalpinaceae, the leaf is bipinnately compound. The primary rachis is modified into a stout spine. This bears the secondary rachis which is modified into a flat phyllode. This phyllode bears the reduced leaflets.

In *Acacia melanoxylon* (Australian acacia) the leaf is bipinnately compound. The primary rachis is modified into a flat phyllode. This bears the secondary rachis with reduced leaflets. When the leaf is mature the leaflets drop down and only the phyllode remains.

In *Barberry*, the leaves on the main stem are reduced to spines. The axillary bud develops into a short branch bearing the normal leaves. The leaves appear as if they are arising on the main stem.

Fig. 7.7 Leaf-Modifications. Tendril in *Naravila* (Central leaflet). Hook in *Bigonia* (Central leaflet)

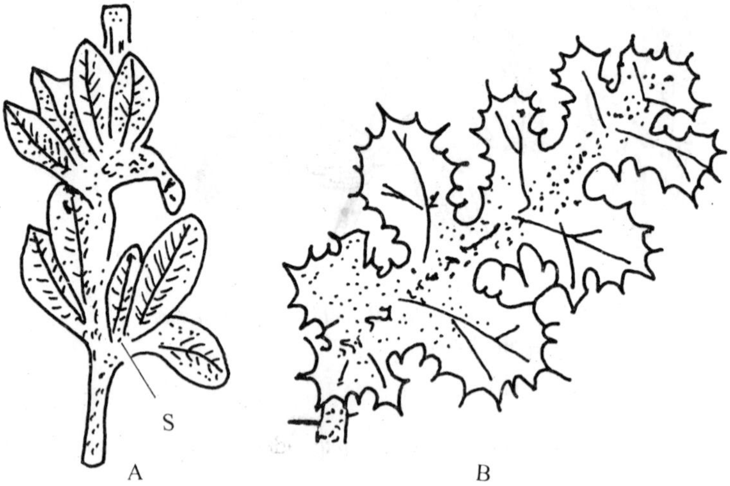

Fig. 7.8 Leaf modifications. **A.** Leaves *Barberry*. **B.** Leaf of *Argemone*

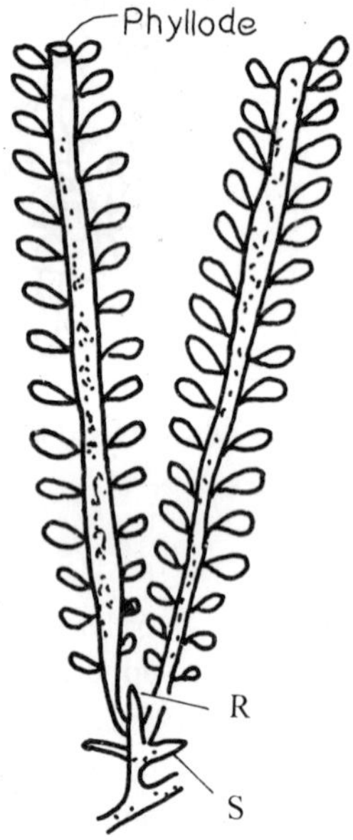

Fig. 7.9 Leaf-Modification in Parkinsonia

In *Opuntia,* the stem becomes a fleshy phylloclade. In order to protect this from grazing animals, the leaves get modified into spines.

Food storage

In many plants the food is stored in the leaves and they constitute the edible parts. Many varieties of greens, lettuse and cabbage are Egamples where food is stored in the leaves.

Reproduction

In *Bryophyllum,* buds arise on the leaves and help in vegetative propagation.

Adaptation to water

Leaves get modified to suit the hydrophytic habitat. In *Eichhornia* the petiole becomes a swollen phyllode, stores air and helps to maintain the buoyancy of the plant.

In *Limnophylla heterophylla* the leaf is of two types. The submerged leaves are highly dissected, while the ones above the water level are broad and entire.

Heterotrophic nutrition

In certain plants growing in marshy areas, the soil will be deficient in nitrogenous contents. Hence the plants (though they are green and can photosynthesise), have an additional means of obtaining their N_2 requirements. In order to meet thier N_2 requirements they trap the insects. The following are some examples.

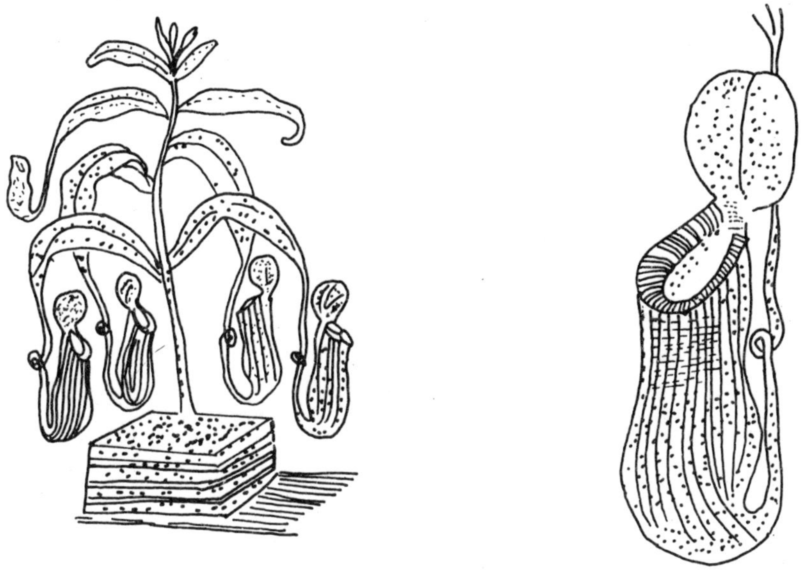

Fig. 7.10 Leaf- Modification (Insectivorous) in Nepenthes

Drosera or Sundew: This plant grows in marshy areas. The leaves have a broad spathulate lamina. On the leaves there are certain gland hairs which secrete a gummy substance. This settles at the tip of the tentacles and glitters like dew drops in sunshine (hence the name sundew plant). Insects are attracted by the glitter of the leaf and when they settle, their body sticks to the leaf. The tentacles bend over the body of the insect and make its escape difficult. Gradually the insects die and the protein part in the body of the insects is digested by the secretion of the enzyme pepsin. The rest of the body is discarded.

Nepenthes : This is commonly called pitcher plant. In our country this grows in the forests of Assam. The plant is a climber. The petiole is modified into a long tendril. The lamina is modified into a hollow pitcher, while the apex is modified inot a lid. The lid is highly colourful and attractive. Inside the pitcher the walls are slippery with hairs pointed downwards. If an insect attracted by the lid happens to fall into the pitcher it cannot escape. Slowly the insect is digested by the leaf.

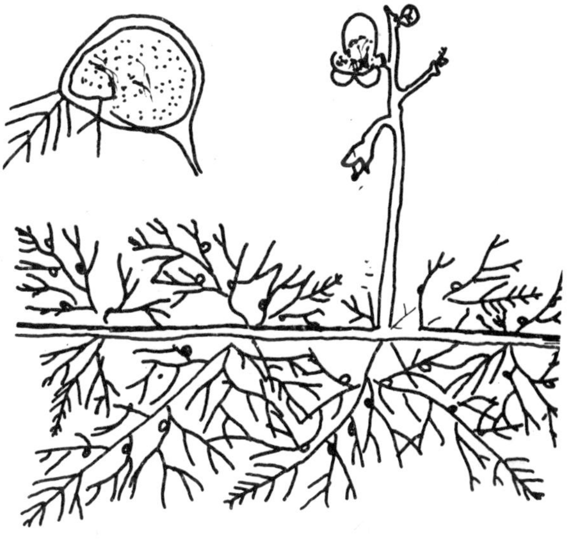

Fig. 11 Leaf- Modification (Insectivorus) in Utricularia

Utricularia : This is commonly called bladder wort and lives in water. There are no roots in this water. There are no roots in this plant. But the leaves are highly segmented and look like roots. Some of the segments are modified into bladders to trap the insects. The bladders to trap the insects. The bladder has a trap door which can open only to the inside. If insects happen to get into the bladder through the water current, they cannot escape. Slowly the insect dies and its body is digested by the leaf.

CHAPTER 8

INFLORESCENCE

INFLORESCENCE

An inflorescence is defined as a group or collection of flowers. In many cases, the individual flowers are inconspicuous and by themselves will not be able to attract the attention of pollinating agents. Under such circumstances they group together and form clusters called inflorescences.

In certain circumstances, when individual flowers are sufficiently large, they are borne singly and said to be solitary.

Types of Inflorescence

Inflorescences are broadly classified into the following types :

1. Racemose
2. Cymose
3. Mixed
4. Special

Racemose types

The following types are recognized under racemose inflorescences.

1. *Simple raceme:* Penduncle is unbranched. Flowers which have stalks are borne directly on the peduncle in acropetalous succession. The tip does not end in a flower eg. *Crotalaria.*

2. *Compound raceme*: Peduncle is branched. Stalked flowers are borne on the branches of the peduncle. A compound raceme is also called a panicle eg. Mango.

3. *Corymb :* Flowers are borne on the short peduncle. There is a difference in the length of stalks of flowers. Older flowers have longer stalks, younger flowers have shorter stalks so that all of them reach almost the same level eg. *Caesalpinia pulcherrima.*

4. *Umbel :*In an umbel inflorescence, the flowers seem to start from the top of the peduncle. All the flowers (old or young) have stalks of similar length so that they reach the same level. In an umbel, there is always a whorl of bracts. An umbel is of two types. Simple and compound. In a simple umbel (eg. *Hydrocotyle*) the peduncle is unbranched. Flowers are borne directly on it. In compound umbel flowers are borne on the branches eg. *Coriandrum.*

Fig. 8.1 Inflorescence – Racemose types **A.** Simple raceme, **B.** Spike, **C.** Spikelet, **D.** Catkin

5. *Spike :* It is a modified raceme; here flowers are sessile (Stalkless) eg. *Achyranthes.* Sessile flowers borne on the branches of the spike constitute a compound spike.

6. *Catkin :*It is a pendulous spike. The peduncle is slender and delicate eg. *Acalypha.* Flowers are unisexual.

7. *Spadix :* It is a spike inflorescence in which the axis is generally fleshy. Further the entire inflorescence is surrounded by a big bract called spathe.

8. *Head or capitulum :* This is an advanced type of racemose inflorescence generally seen in the members of the family Asteraceae. The terminal part of the peduncle is flattened into a receptacle on which the flowers are borne. Two types of heads may be recognized. These are simple and compound. In a simple head (eg. *Tridax*), the peduncle terminates in only one head. In a compound head, the peduncle terminates in several heads. All of them however have a common covering of bracts, besides their individual envelope of bracts eg. *Sphaeranthus, Echinops* etc.

Based on the sex of the florets also, heads may be recognised into two types viz., *Monogamous* and *Heterogamous*. In monogagamous heads, all the florets in the head are alike and all of them are bisexual. In heterogamous heads, there are two types of florets. viz, ray flores and disc florets.

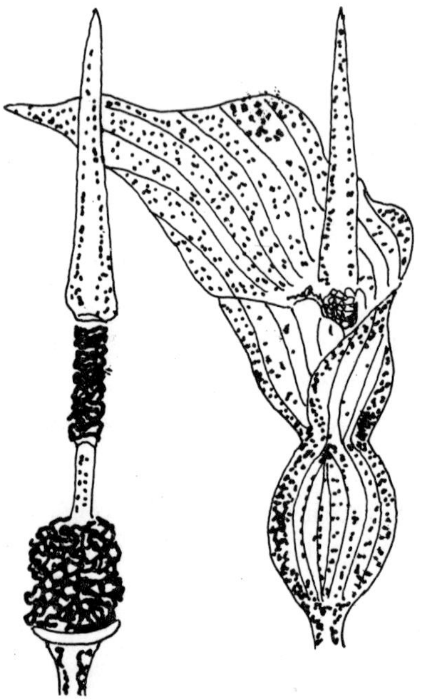

Fig. 8.2 Inflorescence – Spadix Left-without spathe (see the female flowers below) Right-with spathe.

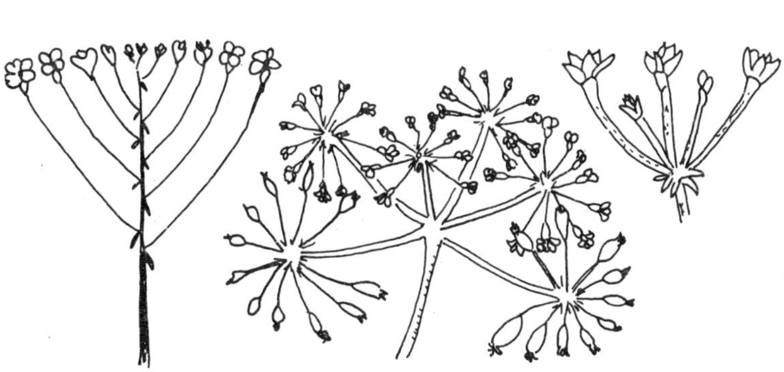

Fig. 8.3 Inflorescence-Spadix Left-without spathe (see the femalf flowers below) Right-with spathe.

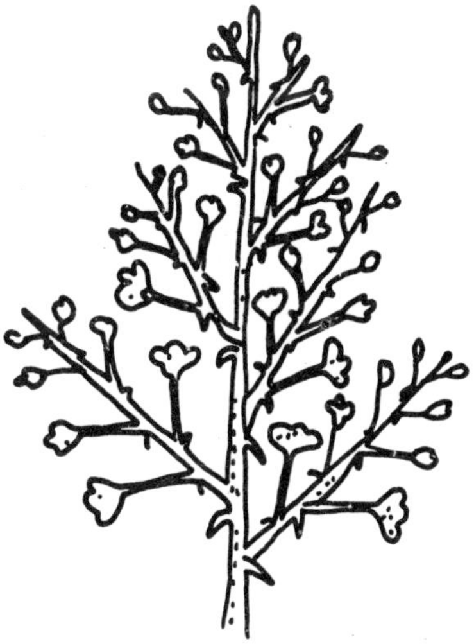

Fig. 8.4 Inflorescence–panicle

Fig. 8.5 Inflorescence – Capitulum
A. Entire, **B.** Long sectional view

Cymose types

The following categories of cymose inflorescences have been recognised-

1 *Simple Cyme* :This consists of only 3 flowers. The central one is the oldest and the two lateral ones are younger eg. *Bougainvillea, Jasmine.*

2. *Monochasial or Uniparous Cyme* :In this type, branching of the peduncle is restricted to only one side and each branch ends in a flower. This is of two types (1) *Helicoid (*eg. *Haemeila*) in which flowers are borne only on one side uniformly throughout the length of the peduncle. (2) *Scropioid* in which flowers are borne alternatively towards left and right eg. *Heliotropium.*

3. *Dichasial cyme or Biparous cyme* :In this type, the tip of the peduncle ends in a flower as in other cymes. The peduncle branches on either side giving rise to two branches. The branches also branch in turn and give rise to branches. All the branches terminate in flowers eg. *Clerodendron.*

4. *Polychasial Cyme or Multiparous Cyme* : The peduncle gives rise to many branches at a point. These branches also in turn behave in the same way. All the branches end in flowers eg. *Nerium.*

Fig. 8.6 Inflorescence – Cymose types.
A. Biparous Cyme, **B.** Scorpioid Cyme, **C.** Helicoid Cyme.

Mixed Types

1. *Thyrusus* :This is a mixture of cymose and racemose types. The peduncle is branched. While the main inflorescence has a racemose development, the branches have cymose development. Here, in the entire inflorescence always older flowers are found at the lower region and the younger ones above. But in each branch older flowers are found towards the apex and the younger ones below eg. *Ocimum*.

Special Types

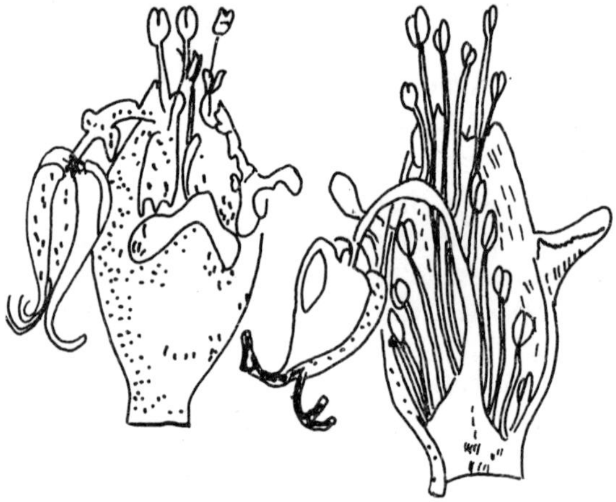

Fig. 8.7 Inflorescence – Cyathium,
A. Entire,**B.** Long sectional view

1. *Cyathium* : This is a special kind of cymose inflorescence, seen in the genus *Euphorbia* (*Poinsettia*). Here the inflorescence is generally surrounded by involucre of five bracts. There may be an extrafloral nectary attached to the bracts. Within the cyathium, there is a single female flower raised on a stalk called *Gynophore*. Surrounding the single female flower, there are five groups of male inflorescences each arranged in a scorpioid cyme. Both male and female flowers are achlamydeous in the sense that they do not have either calyx or corolla.

2. *Hypanthodium* : This inflorescence is seen in *Ficus*. The inflorescence is a cup-like structure having an opening called the ostiole. Lining the inner wall of this pear-shaped structure are found male and female flowers.

The receptacle is formed by the condensation of rachis of three closely placed cymes. The unisexual flowers are arranged in cymose group.

3. *Verticellaster* : It is a special type of cymose inflorescence found in the members of the family Labiatae (eg. *Leucas*, etc). Here, at the nodal region, there are two dichasial cymes. In these cymes, only the first branching is dichasial, while all the further branches are monochasial (helicoid). Such a type of combination of dichasial and monochasial cyme is called verticellaster. The word verticellaster means false whorl. Since the flowers are sessile, they form a false whorl around the nodal region.

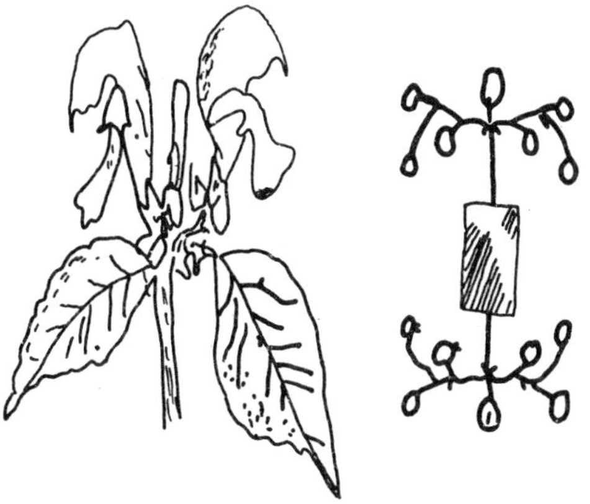

Fig. 8.8 Inflorescence – Verticellaster
A. Inflorescence of Leucas, **B.** Diagrammatic representation of the Inflorescence.

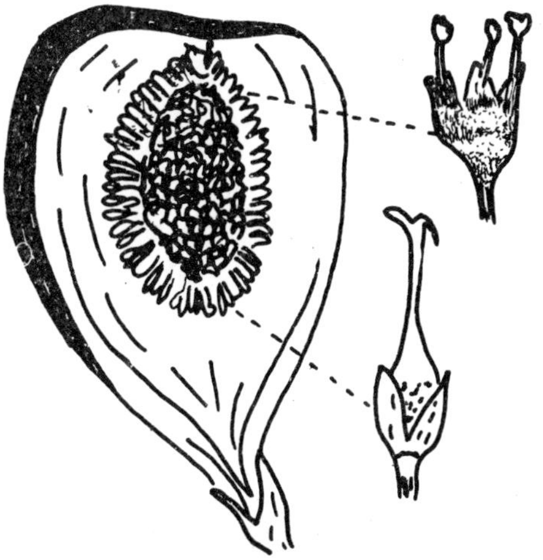

Fig. 8.9 Inflorescence–Hypanthodium. Male (above) and female (Below) flowers are enlarged at the right side.

CHAPTER 9

FLOWER

FLOWER

A flower is a shoot modified for reproduction . It is recognised as the modified vegetative shoot. A flower is also supposed to possess nodes and internodes and leaves at the nodal regions just like a vegetative axis. But in flower, unlike in a vegetative shoot the internodal regions are condensed, so that successive whorls of leaves appear clustered. The leaves present in a flower are called floral leaves. Most of the times, the floral leaves do not resemble the normal leaves as their function is different. The main function of the floral leaves is reproduction.

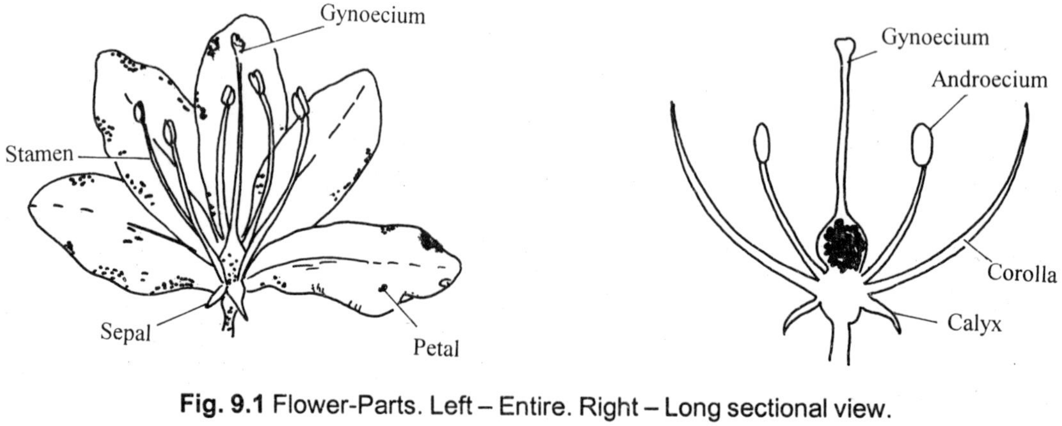

Fig. 9.1 Flower-Parts. Left – Entire. Right – Long sectional view.

Parts of a Typical Flower

Basically a flower has two types of organs viz, accessory organs and essential organs. Accessory organs are those which assist but do not directly take part in reproduction. Essential organs on the other hand directly take part in reproduction. In some cases, accessory organs may be missing in a flower. But essential organs are always present.

Accessory Organs

Generally there are two types of accessory organs. These are calyx and corolla. Calyx is usually greenish in colour, while corolla is variously coloured. Calyx serves to protect the flower in bud condition. Corolla serves to attract the polinating agents. The individual members of calyx are called sepals, while those of corolla are called petals.

Essential Organs

These are two types-androceium and gynoecium. Andorecium is the male reproductive structure and gynoecium is the female reproductive structure.

Evidences for flower being a modified shoot

(a) Basic construction of the flower is on the same lines of a shoot. It is not exactly similar to a vegetative shoot because its functions are different.

(b) Foliar nature of floral organs : The floral organs, at least the calyx retains the foliar nature. The individual members are greenish in colour and photosynthetic.

(c) In some cases, transitional types occur bridging the gap from the vegetative shoot to the flower. Sometimes stamens, petals, and sepals appear like leaves. This is very common in *Nymphaea* (Water lily).

(d) In certain roses, from the centre of the flower a vegetative shoot arises, giving support to the foliar nature of a flower. This is an instance of the continuation of the growth of the axis.

Fig. 9.2 Flower– Abnormal growth of vegetative shoot from the centre of the flower in Rose.

Detailed Description of terms used in Describing the flower

A flower is usually stalked, if it is having a stalk it is said to be *pedicellate*. If there is no stalk it is said to be *sessile.* The pedicel terminates in a flattened structure called the receptacle or the thalamus. The thalamus may be domeshaped, cup-like or flattened.

Bracteate **:** Bract present.

Ebracteate : Bract absent.

Bract is a leafy appendage, present below the pedicel and usually on the stem. It protects the flower in the young condition.

Bracteole is another leafy appendage. But this is present on the pedicel and below the calyx.

Bracteolate **:** Bracteole present.

Ebracteolate **:** Bracteole absent.

Complete **:** When a flower has all the organs (both accessory and essential).

Incomplete **:** Any one of the floral organs is missing.

Homochalmydeous **:** Accessory organs are present, but cannot be distinguished into calyx and corolla. In such cases, the members of accessory whorl are called perianth lobes or tepals e.g. *Michelia champaca.*

Achalmydeous **:** Accessory organs are completely absent, only essential organs are present e.g. *Poinsettia.*

Heterochlamydeous : Accessory organs present and can be distinguished into calyx and corolla.

Phyllotaxy of floral leaves **:** This explains the manner of arrangement of the floral leaves on the receptacle. There are 3 types-

(a) *Acyclic* : Floral leaves arranged in a spiral manner on the elongated receptacle e.g. *Magnolia.*

(b) *Cyclic or Whorled* : Floral leaves are arranged in definite circles or whorls e.g. *Achanthus.*

(c) *Hemicyclic* : Some members are cyclic and some acyclic e.g. *Anona.*

Sex of the Flower

Bisexual or Hermaphrodite : Both the sexes (Andorecium and Gynoecium) are present in the flower.

Unisexual : Only one sex is present in the flower. Female flowers are called pistillate, while male flowers are called staminate.

Distribution of flowers of different sexes in a plant

(a) *Monoecious* : Flowers unisexual. But both male and female flowers present in the same plant e.g. *coconut.*

(b) *Dioecious* : Flowers unisexual and distributed in different plants e.g. *Vallisneria.*

(c) *Polygamous* : When the plant consists of 3 types of flowers (a) bisexual, (b) male and (c) female e.g. *Mango.*

Position of the Gynoecium

The position of the gynoecium with reference to other floral parts is an important criterion in the description of the flower. Based on this, flowers are of three types.

(a) *Hypogynous* : All the floral parts are arranged below the gynoecium. Hence gynoecium is superior. In hypogynous flowers the receptacle is dome-like and all floral parts arise on it e.g. Members of Leguminosae.

(b) *Perigynous* : Here all the floral parts arise above the level of gynoecium. But the gynoecium is on the receptable which is cuplike. Hence gynoecium is neither superior nor inferior. It is said to be semi inferior e.g. Members of Rosaceae.

(c) *Epigynous* : All the floral parts arise above the gynoecium. The gynoecium is sunk into the receptacle and the wall is inseparable from the wall of the latter. Here the gynoecium is clearly inferior ex. *Members* of *Asteraceae.*

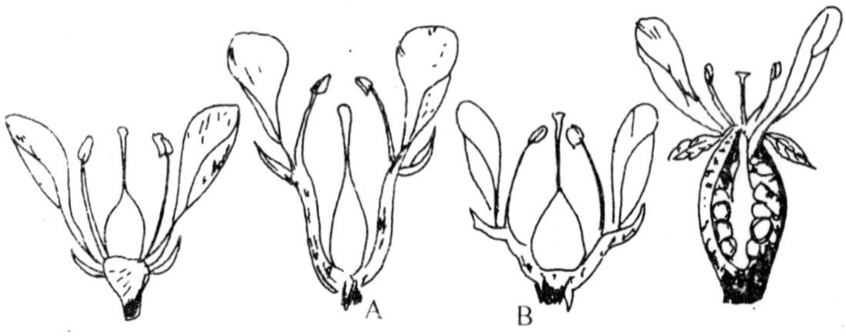

Fig. 9.3 Flower – Positions of ovary (from the left) hypogynous, perigynous (two diagrams) and epigynous.

DETAILS OF FLORAL PARTS

Calyx

The individual members are called sepals. The sepals in the whorl of calyx may be free (polysepalous) or fused (gamosopalous, synsepalous).

Corolla

The members are called petals. They may be free (polypetalous) or fused (sympetalous or gamopetalous).

The main function of the corolla is to attract the pollinating agents.

Aestivation

The arrangement of petals or sepals in the floral bud (flower) is called aestivation. Chiefly the following types are seen.

(a) *Valvate* : In this type, the members of a whorl are in contact with each other by their margins, i.e. all the petals or sepals are adjacent e.g. *Anona. Calotropis.*

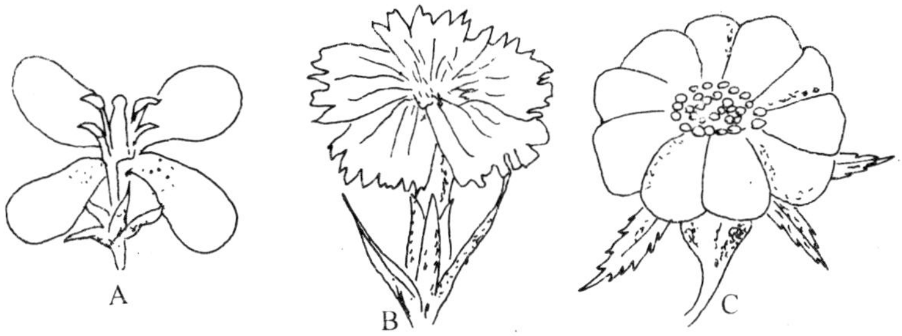

Fig. 9.4 Flowr–types of corolla **A.** Cruciform, **B.** Caryophyllaceous, **C.** Rosaceous

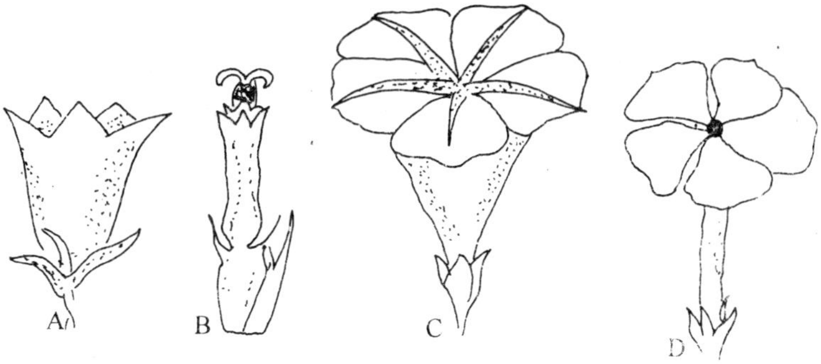

Fig. 9.5 Flower–types of Corolla **A.** Campanulate, **B.** Tubular, **C.** Funnel shaped, **D.** Rotate

(b) *Imbricate* : The sepals or petals overlap. Usually one petal or sepal is completely out, its both margins are overlapped, another one is completely external. The remaining three petals or sepals have one margin outside (overlapping) and the other margin inside (being overlapped). Two types of imbricate aestivation are distinguished.

(i) *Ascendingly imbricate* : The odd petal or sepal is innermost e.g. *Cassia.*

(ii) *Descendingly imbricate* : The odd petal or speal is outermost e.g. *Pea.*

(c) *Twisted or Contorted* : One of the margins of each petal is outside and the other margin inside i.e. one of the margin of a petal or sepal overlaps and the other margin is being overlapped. Thus the petals or sepals look like twisted, e.g. *Hibiscus.*

(d) *Quincuncial* : In this type, 2 petals or sepals lie completely outside and the other 2 lie inside, remaining fifth petal has one margin outside and other margin inside. e.g. *Flowers of Scrophulariaceae (Digitalis).*

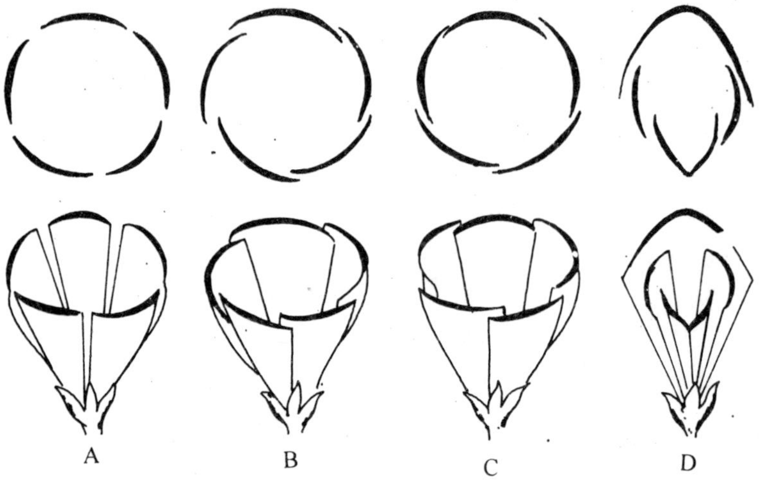

Fig. 9.6 Flower–Types of aestivation
A. Valvate, **B.** Twisted, **C.** Imbricate, **D.** Ascendingly imbricate **C.**Vexillary

Androecium

This is the whorl of male reproductive structures. The individual members are called stamens. Each stamen has 3 parts viz., filament, anther and connective. Filament is the stalk-like structure. Anther is the terminal portion in which the pollen grains are formed. Connective is the portion that joins the anther and the filament. The following are some of the terms used in describing the androecium.

Monadelphous : All the stamens (filament) in a flower unite to form one bundle. e.g. : *Hibiscus.*

Diadelphous : Stamens join together to form two bundles e.g. : *Pea, Bean.*

Polyadelphous : Stamens group together to form many bundles e.g. lemon.

Syngenesious : This is a condition in which all the anthers are fused but filaments are free. e.g. Members of Asteraceae.

Synandrous : Stamens are united completely by both filaments and anthers eg : Cucumber.

Monothecous : Only one lobe in the anther eg: *Hibiscus.*

Dithecous : Two lobes in the anther eg: *Cassia.*

Staminode : A sterile stamen.

Didynamous : Four stamens present. Two tall and two are dwarf.

Tetradynamous : Six stamens present of which four are long and two are dwarf (short)

Attachment of Anther to the Filament : Mainly there are four types of attachment.

(1) *Basifixed* : Filament attached to the base of the anther eg. Radish

(2) *Dorsifixed* : The tip of the filament is attached to the back of the anther (about the middle region) eg. passion flower

(3) *Adnate* : The filament is fully attached to the anther eg. *Michelia*

(4) *Versatile* : The filament is attached to the anther at the middle, so that the anther is loose and can oscillate. eg. grass.

Gyneocium

This is also called the pistil. There are three parts in this viz., ovary, style and stigma. Stigma is the terminal portion. It serves as a platform for the lodging of the pollen grains. Style is a long stalk like structure through which the pollen tube grows. Ovary is the basal portion of the pistil. It may be superior, inferior or semi-inferior. It is made up of carpels. The following terms are used in the description of the ovary.

Monocarpellary	:	Ovary is made up of one carpel
Bicarpellary	:	Ovary is made up of two carpels
Tricarpellary	:	Ovary is made up of three carpels
Multicarpellary	:	Ovary is mad up of many carpels
Syncarpous	:	Carpels tow or more in number and fused.
Apocarpous	:	Carpels tow or more and free.

The ovary is divided into one to many chambers called locules. Ovary may have one (monolocular), two (biolocular), three (trilocular) or many (multilocular) locules.

Placentation : The ovules are attached to the ovarian tissue through a tissue called placenta. The type of attachment is called placentation. Chiefly, the following types of placentation are seen.

1. *Axile* : There is a central axis running in the centre of the ovary. The ovules are attached to it. Ovary is divided into many chambers (as many as carpels) by the septa radiating from the central axis. e.g. Malvaceae.
2. *Parietal* : The ovules are attached to the fused inner margin in many groups, the ovary is one chamberad e.g. Papavaraceae.

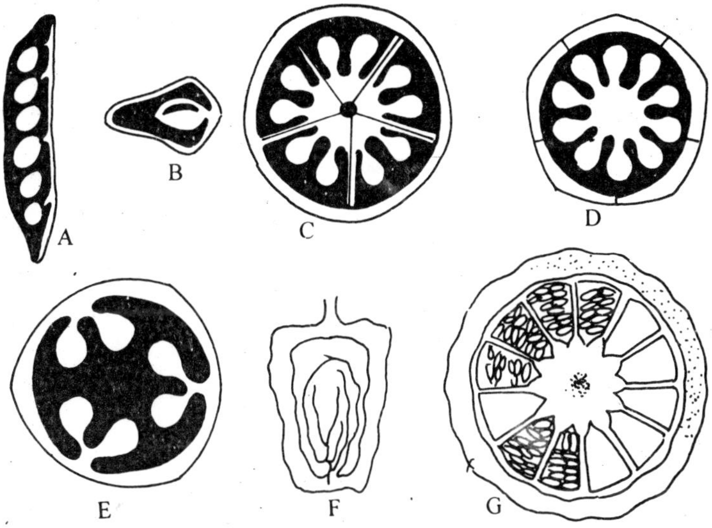

Fig. 9.7 Types of placentation **A.** Marginal (L.S), **B.** Marginal (T.S.), **C.** Axile (T.S.) **D.** Free central (T.S.), **E.** Parietal (T.S.), **F.** Basal (L.S.), **G.** Superficial (T.S.)

3. *Marginal* : The ovules are attached to the margin in one row (group). The ovary here is usually monocarpellary e.g. Leguminosae.
4. *Basal* : Usually the ovary is bicarpellary and unilocular. There is only one ovule arising from the floor of the ovary e.g. Asteraceae.
5. *Free Central* : The centre of the ovule has a hollow mound on which are arranged the ovules. The ovules are free from the ovary wall. Axile placentation by the loss of septa becomes free central type e.g. Primrose.

CHAPTER 10

FRUITS

FRUITS

A fruit may be defined as a fertilised (mature) ovary. After fertilization an ovule becomes a seed, and the ovary swells up and becomes the fruit.

In some cases, however, the ovary may ripen and develop into a fruit without fertilization. Such a phenomenon is said to be '*Parthenocarpy*'. Banana is a very good example of parthenocarpy. There are no seeds in the fruit.

Classification of Fruits

Basically fruits are classified into (1) True fruits and false fruits. True fruits are those where the ovary develops into the fruit. In a false fruit, also called the *pseudocarp*, any part of the flower, other than the ovary, after fertilization becomes swollen and looks like a fruit. Apple and cashew fruits are typical false fruits. In apple the fruit is called a pome. Here the thalamus or the receptacle becomes flesly and edilble.

Fruits are classified into the following categories :

1. Simple fruits.
2. Aggregate fruits, and
3. Multiple or compound fruits.

Simple Fruits

A simple fruit is one that develops from a monocarpellary or multicarpellary ovary of a single flower. In other words only one fruit develops (separately) from one ovary.

Simple fruits are again classified into two types, based on the nature of the fruit wall (pericarp). These are the *fleshy fruits* and *dry fruits*. In fleshy fruits, the fruit is juicy and fleshy. These are indehiscent. The fruit wall is generally distinguishable into three layers namely the outer *epicarp*, the middle *mesocarp* and the inner *endocarp*. In dry fruits, the fruit wall cannot be distinguished into three layers.

Fleshy fruits are of two types-Baccate and Drupaceous. In baccate fruits all parts of the fruit are soft except the seeds. In drupaceous fruits however, the endocarp is hard and stony. Baccate or fleshy fruits are of the following types-

1. *Berry or Becca :* (eg. Tomato, Banana)

Here the fruit is fleshy and juicy. The epicarp is thin. The mesocarp and sometimes even the endocarp is fleshy in which the seeds are embedded.

2. *Hesperidium :* eg. Citrus fruits like lime, lemon, pomello etc.

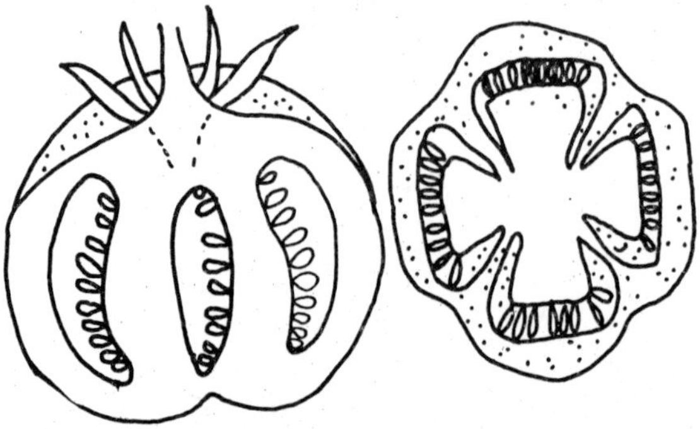

Fig. 10.1 Fleshy Fruits
Berry long section (left) cross section (right)

This is a type of berry. This is characteristically found in the members of the family Rutaceae. Here the epicarp form the outer glandular rind; while the mesocarp forms the white fibrous structures attached to the inner surface of the epicarp. The endocarp constitutes the inner portions. It forms septa which converge towards the center.

3. *Pome :* (eg. Apple)

This is actually a false fruit. Here the soft edible portion is the receptacle or thalamus. The seed-like structures that are discarded are the actual fruits. (Achenes).

4. *Drupe :*(eg. Cocoanut, Mango)

Here the fruit has a thin epicarp. The mesocarp in generally fibrous. It may be even fleshy as in mango. The endocarp is very hard. The seed is enclosed in the endocarp. In mango, the edible portion of the fruit is the mesocarp. In cocoanut, the endosperm (both liquid as well solid) is the edible portion and the mesocarp is fibrous.

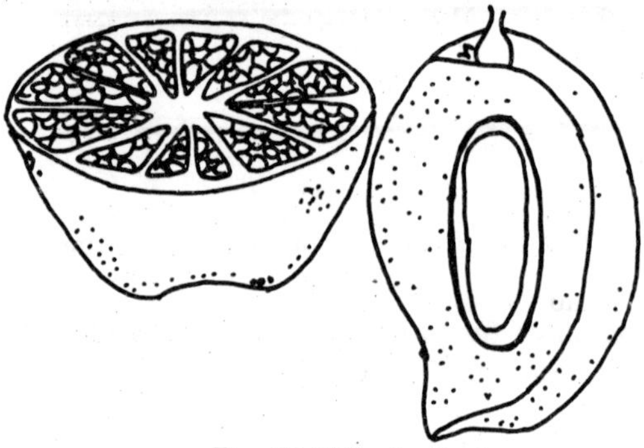

Fig. 10.2 Fleshy Fruits
Hesperidum (left) Drupe of Mango

5. *Pepo:*

This is a type of berry developing from an inferior ovary. This is found in the family Cucurbitaceae (eg. pumpkins, cucumber etc). In this fruit, the epicarp and thalamus form the outer hard leathery rind. The mesocarp, endocarp and the placenta constitute the pulp and are edible.

Dry fruits

These are further classified into two types viz., dry dehiscent and dry indehiscent. In dry dehiscent fruits, the fruits break open at maturity exposing the seeds. In dry indehiscent fruits, the fruits do not open even at maturity. In this case the fruits themselves are dispersed.

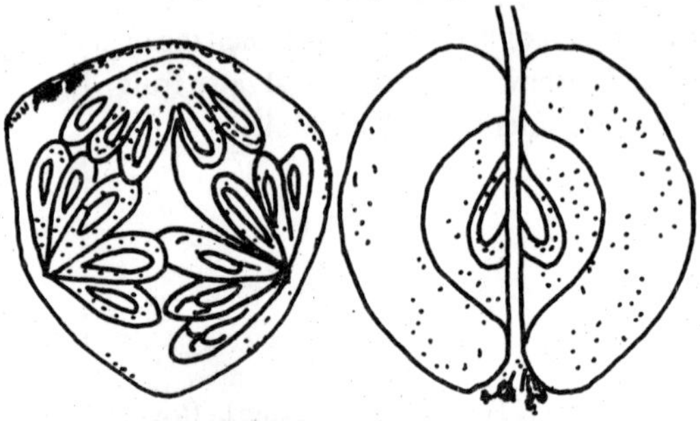

Fig. 10.3 Fleshy Fruits– Pepo in cross section (left) Pome in long section (right)

(a) Dry Dehiscent fruits

(1) *Legume :* This is characteristic of the family Leguminosae. The fruit developing from a monocarpellary superior ovary encloses many seeds on marginal placenta. The fruit breaks open on its ventral wall exposing the seeds (eg. *Pisum).*

Pod and lomentum are variations of legume. A few seeded lugume (eg. *Arachis*-ground nut) is called a pod, whereas a deeply chambered legume is called a lomentum (eg. Soap nut).

(2) *Follicle :* A follicle is essentially like a legume, but differs from it in splitting along only one suture. The fruit develops from a superior, monocarpellary unilocular ovary eg. *Delphinium.*

(3) *Siliqua :* The fruit develops from a bicarpellary ovary. The ovary which is bilocular in the beginning becomes tetralocular due to the development of a false septum called *replum.* The fruit opens on both the valves with the seeds attached to the central replum.

(4) *Capsule :* A capsule is a dry dehiscent fruit developing from a bi or multi-carpellary ovary with seeds attached to the fruit on an axile placenta. The fruit breaks open in two different ways. In one type called

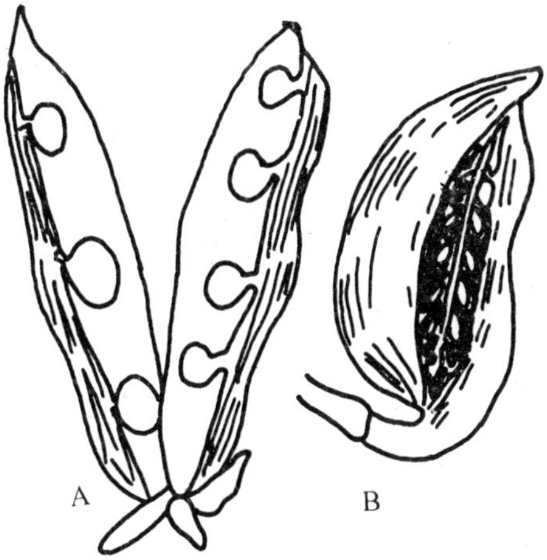

Fig. 10.4 Dry dehiscent fruits
A. Legume, **B.** Follicle

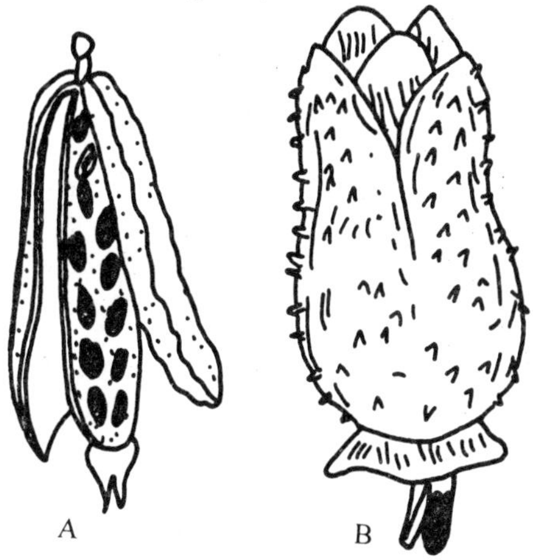

Fig. 10.5 Dry dehiscent fruits
A. Siliqua, **B.** Capsule.

septicidal, the fruit breaks open along the septa dividing the locules. In the other type called *loculicidal*, the fruit splits open in the middle of a locule, in between the two septa.

There is a third type of splitting in the capsule called *septifragal* (eg. *Datura)*. Here the fruit may break open either along the septa or in between them, but the septa break from the outer wall of carpels with the seeds attached to the centre.

Septicdial capsule eg. *Colchicum*

Loculicidal capsule eg. *Gossypuim, Hibiscus* etc.

(b) Dry Indehiscent fruits

(1) *Achene :* It generally develops from a monocarpellary ovary and has only one seed. The pericarp is hard and leathery eg. *Nelumbium.*

(2) *Caryopsis :* It is an achene in which the seed coat and the fruit wall are fused and cannot be separated eg. Maize, Paddy etc.

(3) *Cvypsela :* The fruit is an achene, but it develops from an inferior ovary. The ovary is usually bicarpellary, syncarpous with a single basal ovule eg. fruits of Asteraceae.

(4) *Samara :* This is a large winged achene fruit. The wings are formed by the expansions of the pericarp e.g *Dipterocarpus, Hiptage, Maple.*

(5) *Nut :* It is a dry indehiscent fruit covered by hard and woody pericarp. It develops from bi or polycarpellary ovary eg. Cashewnut and Chestnut.

Schizocarpic fruits

This is the general name given to the fruits, which combine the characters of both dry dehiscent and dry indehiscent fruits. Here the main fruit splits open (*schizo* = to split), but the

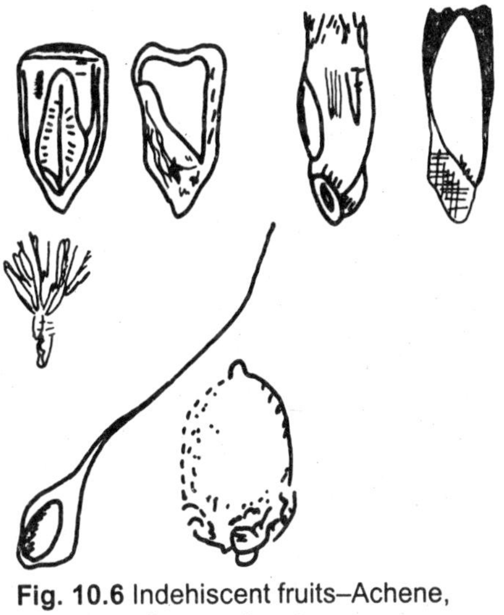

Fig. 10.6 Indehiscent fruits–Achene, Caryopsis, etc.

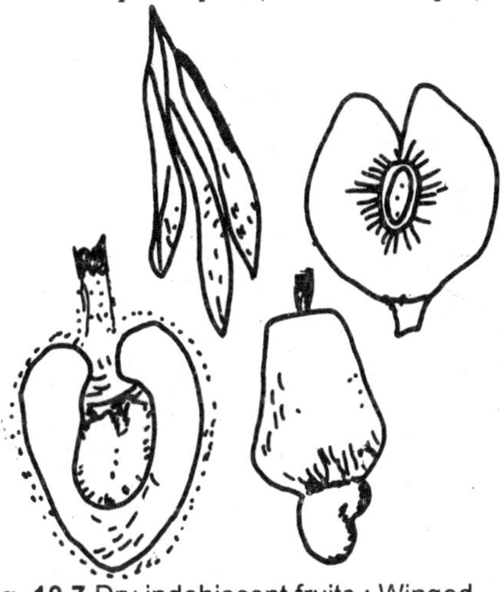

Fig. 10.7 Dry indehiscent fruits : Winged fruits and Nut. (Anacardium

seeds are still trapped inside the locules, so that even though the fruit splits open, the seeds are not set free. The fruit breaks up into many bits enclosing the seeds. Schizocarpic fruits are found in Malvaceae (*Abutilon),* Umbelliferae (*Coriandrum)* etc.

Schizocarpic fruits are of 4 types

1. *Lomentum* : It is similar to legume but does not open. The fruit is partiitoned into number of one seeded compartments eg. *Acacia.*

2. *Cremocarp* : It splits into 2 mericarps which are one seeded pieces. Mericaprs are attached to an axis called caropphore. Generally it develops from inferior bicarpellary ovary eg. *Coriander, Cumin.*

3. *Regma* : It developes from tri or polycarpellary syncarpous ovary. Fruit splits away from the central axis into 3 or many parts called cocci eg. *Castor, Euphorbia.*

4. *Carcerule* : It develops from bi, or multicarpellary ovary. The fruit splits into 4 or many mericarps. Generally persistent calyx are present as in fruits of Labiatae. Other examples are *Abutilon* and *Malva.*

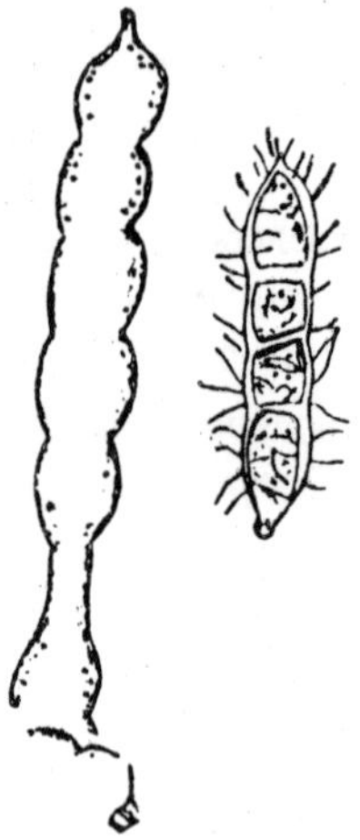

Fig. 10.8 Schizocarpic fruits – Lomentum

Fig. 10.9 Schizocarpic fruits – Double Samara

Aggregate Fruits

Aggregate fruits develop from a multicarpellary, apocarpous ovary of a single flower. Each carpel develops into a small fruit or fruitlet. The collection of fruitlets is known as *etaerio.* Each fruitlet in the aggregate may be an achene, berry or drupe or follicle.

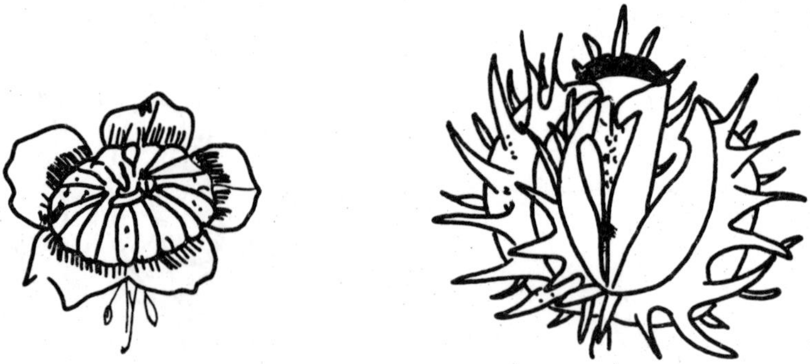

Fig. 10.10 Schizocorpic fruits : Left–Carcerula; Right–Regma

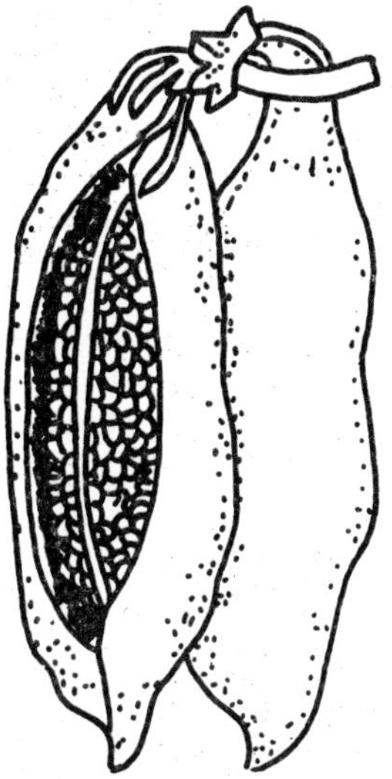

Fig. 10.11 Aggregate fruits; etaerio of Follicles.

(1) *Aggregate of Follicles* : This is found in the family Magnoliaceae. Here each fruitlet is follicle, which is a dry dehiscent fruit. The fruitlet dehisces from the ventral suture only.

(2) *Etaerio of Achenes* : eg. rose, strawberry and lotus.

(3) *Etaerio of Drupes* : Small drupes or drupelets form an aggregation eg. Raspberry.

(4) *Etaerio of Berries* : Here berries may lie embedded in fleshy thalamus as in custard apple (Anona) or remain separate as in *Polyalthia*.

Multiple or Compound Fruits

In this type of fruit, the entire inflorescence develops into a single fruit. There is a common envelope for the entire fruit.

1. Sorosis : It is found in *Mulberry, Pineapple* and *Jack fruit*. These are edible fruits that are derived from catkin (*Mulberry*), spike (*Pineapple*) and spadix (*Monstera deliciosa*) inflorescences.

In *Morus (Mulberry)* the female catkin develops into a sorosis. The female flowers in the catkin have 4 free perianth lobes and two united carples that are unilocular and have one basal ovule. The four perianth lobes become thick and juicy and are edible. The juicy perianths of individual flowers arrange themselves compactly and the whole composite fruit is edible.

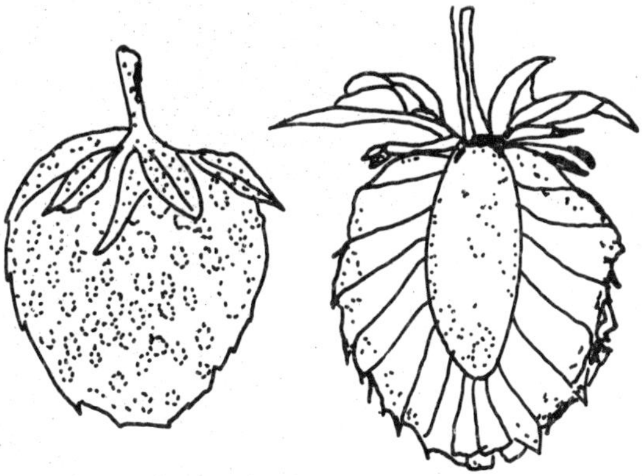

Fig. 10.12 Aggregate fryuts– Etaerio of drupes, left entire, right long section

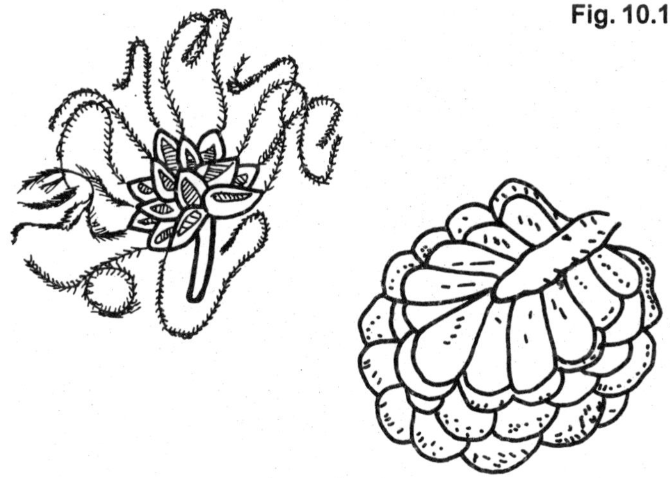

Fig. 10.13 Aggregate fruits : Etaerio of achenes (left) Etaerio of berries (right)

In *Pineapple*, the inflorescence axis or the peduncle bears sessile flowers arranged in a spike. The apex of the spike continues to grow and produces a crown of leaves. The individual flowers in the spike are joined together from the time they are first formed. The flowers are bracteate and are bisexual with distinct sepals and petals (3 each). The ovaries posses abortive ovules that fail to develop into seeds in the cultivated varieties. The peduncle of the inflorescence becomes thick and juicy.

Due to excessive growth of the peduncle the flowers become embedded in it. The petals and bracts of the flower project beyond the surface of the thick peduncle. As the inflorescence matures it becomes more juicy, the sepals, petals and styles and stamens wither away. The bract of each flower, no doubt, projects beyond the ripe fruit and is persistent. The rhomboid markings on the surface of the fruit represent the surface area of each flower and its position on the peduncle. The bract is seen as a small projection The carples in some wild species of *Pineapple* develop into berries which are embedded in the fleshy axis.

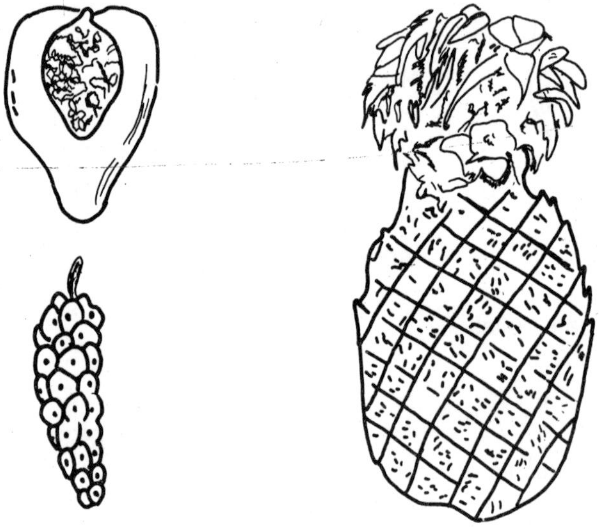

Fig. 10.14 Multiple fruits : Syconus (Top) Sorosis of Pineapple (right). Sorosis of Mulberry (below)

2. **Syconus :** It is a fruit characteristic of the *genus Ficus* and is derived from the entire inflorescence called the **hypanthodium**. In this case the deep cup-shaped receptacle becomes thick, fleshy and juicy. It is the edible part. There are three types of flowers - male, femlae and gall flowers. The female flowers with long styles develop into nut-like fruits, where as the others remain sterile or wither. The opening of the fruit is guarded by scales. The edible fig is *Ficus carica.* The syconuses of Peepal and Banyan are eaten by birds.

In Jack fruit, within the compound fruit the perianth lobes are fleshy and are edible.

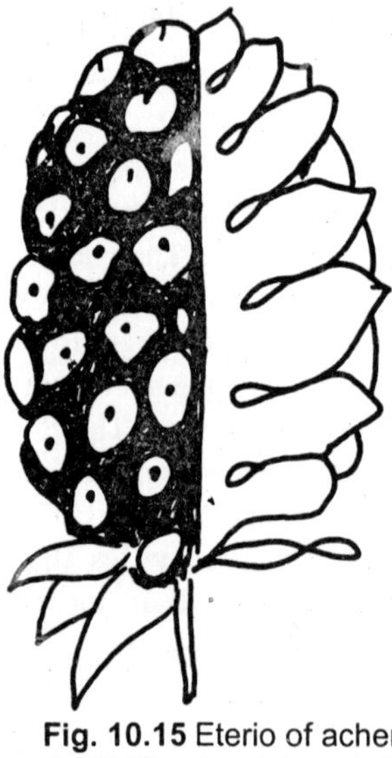

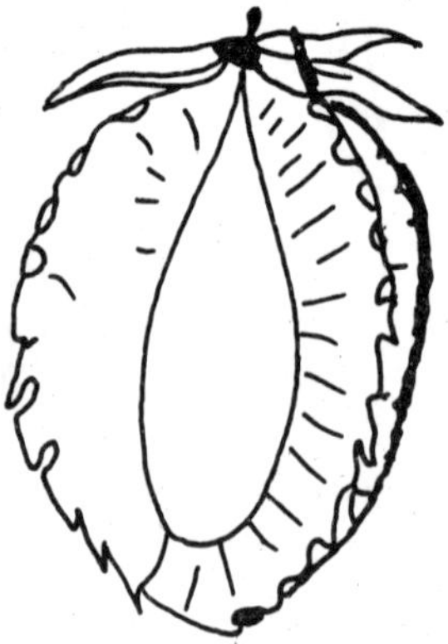

Fig. 10.15 Eterio of achenes of straw berry (Left) Etaerio of drupelets of Rubus (Right)

CHAPTER 11

STUDY OF PLANT FAMILIES

Family : MALVACEAE (Eg : *Hibiscus rosa sinensis*)

Habit : A perennial shrub; ornamental, cultivated in gardens.

Root : Tap root system

Stem : Aerial, erect, woody, cylindrical and glabrous, often green in colour.

Leaf : Alternate, petiolate, simple, stipulate and ovate with serrate margin. Venation is reticulate and multicostate.

Inflorescence : Solitary, axillary or terminal.

Flower : Pedicellate, ebractate, complete, cyclic, actinomorphic, pentamerous, bisexual and hypogynous; epicalyx present outside the calyx.

Epicalyx : 6-8 in number and free, forming a whorl outside the calyx.

Clayx : Sepals 5, gamosepalous; aestivation valvate.

Corolla : Petals the, polypetalous, fused at the base with the staminal tube; red in colour and aestivation twisted.

Androecium : Stamens indefinite in number, monadelphous and filaments united to form a staminal tube surrounding the gynoecium. Anthers monothecous, dorsifixed and reniform the shape. Staminal column attached to the petals at the base.

Gynoecium : Ovary superior, multicarpellary, syncarpous and multilocular with many ovules in each locule on axile placenta. Style long and slender, stigma 5 in number.

Fruit : A capsule

Floral formula : $E_{6\text{-}8}\ K_{(5)}\ C_5\ A_{\infty}\ G_{(5)}$

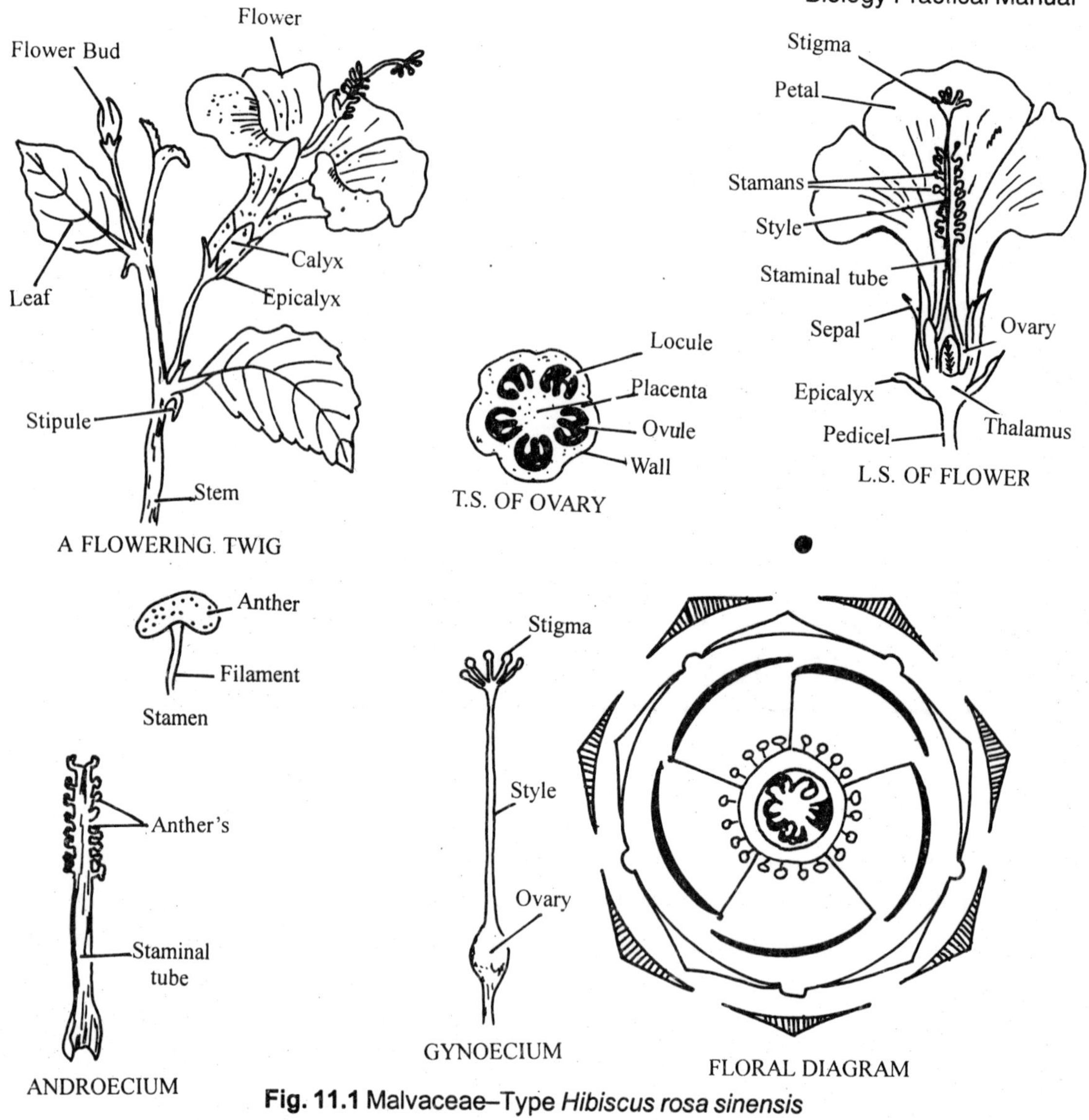

Fig. 11.1 Malvaceae–Type *Hibiscus rosa sinensis*

Family : SOLANACEAE (Eg, *Datura stramonium*)

Habit : A wild annular herb; poisonous.

Root : Tap root.**Stem :** Herbaceous, aerial, erect, cylindrical, green. Slightly sistular (hollow).

Leaf : Alternate, simple, petiolate, exstipulate, ovate; margin entire; apex acute, slightly hairy, venation unicostate and reticulate. Leaf base unequal.

Inflorescence : Flowers large, white coloured, solitary and axillary.

Flower : Pedicellate, ebracteate, ebracteolate, complete, pentamerous, cyclic, bisexual and hypogynous.

Calyx : Sepals 5 in number, gamosepalous; aestivation twisted; clayx persistent even in the fruit.

Corolla : Petals 5, gamopetalous white in colour, campanulate; aestivation twisted.

Andorecium : Stamens 5, free, epipetalous, alternate with the petals; filaments unequal; anthers dithecous and basifixed.

Gynoecium : Ovary superior, bicarpellary, syncarpous, bilocular at the tip, but becomes tetralocular at the base due to the development of a false septum. Ovules many on axile placenta. Carpels obliquely placed. Long style and bilabial stigma.

Fruit : a septifragal capsule.

Floral formula : $K_{(5)} C_{(5)} A_5 G_{(2)}$

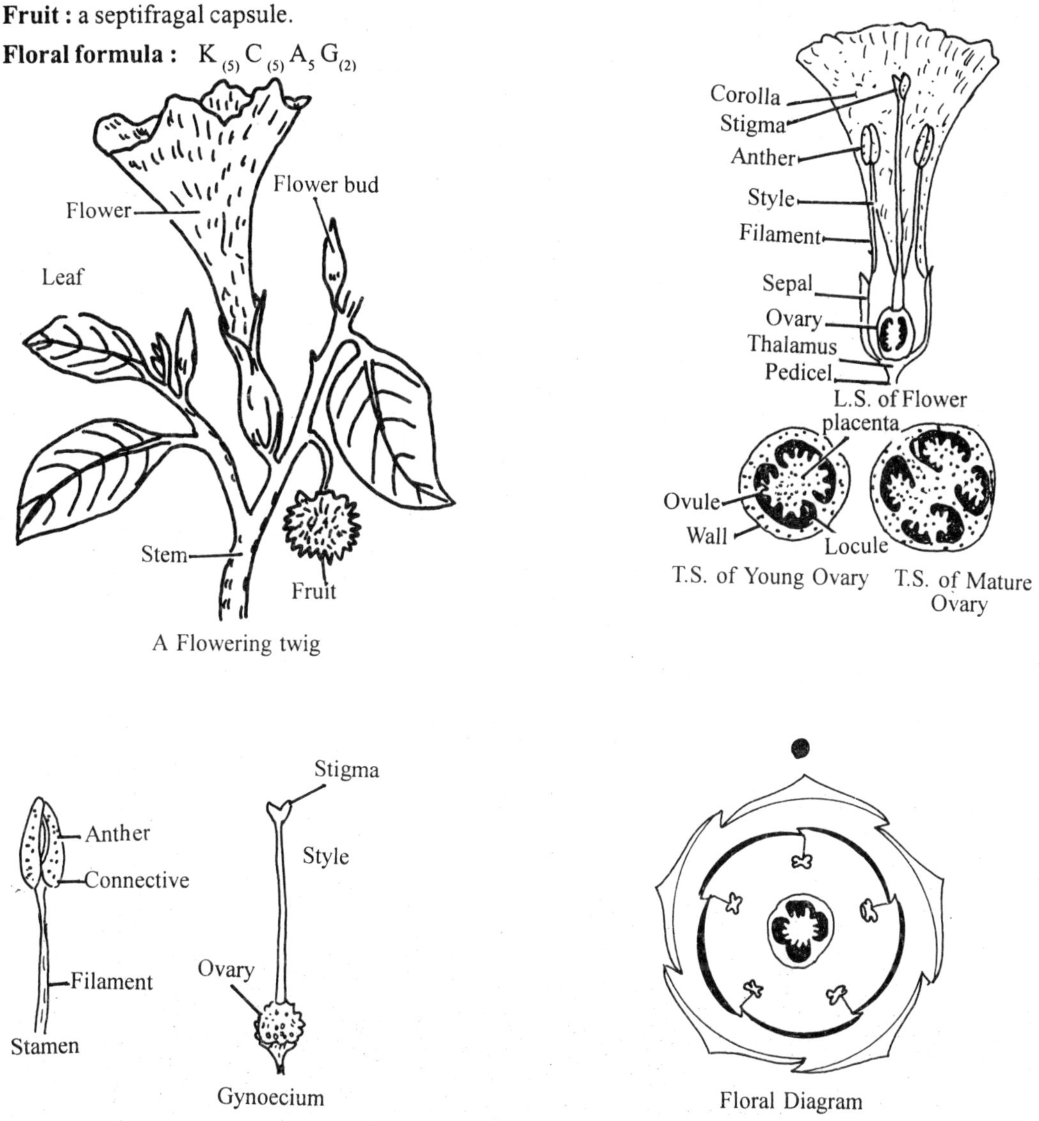

Fig. 11.2 Solanaceae– Type *Datura stramonium*

Family : PAPILIONACEAE (Eg. *Crotalaria*)

Habit : A wild herbaceous plant.

Root : Tap root; bacterial root nodules present.

Stem : Aerial, erect, cylindrical, green in colour, branched and herbaceous.

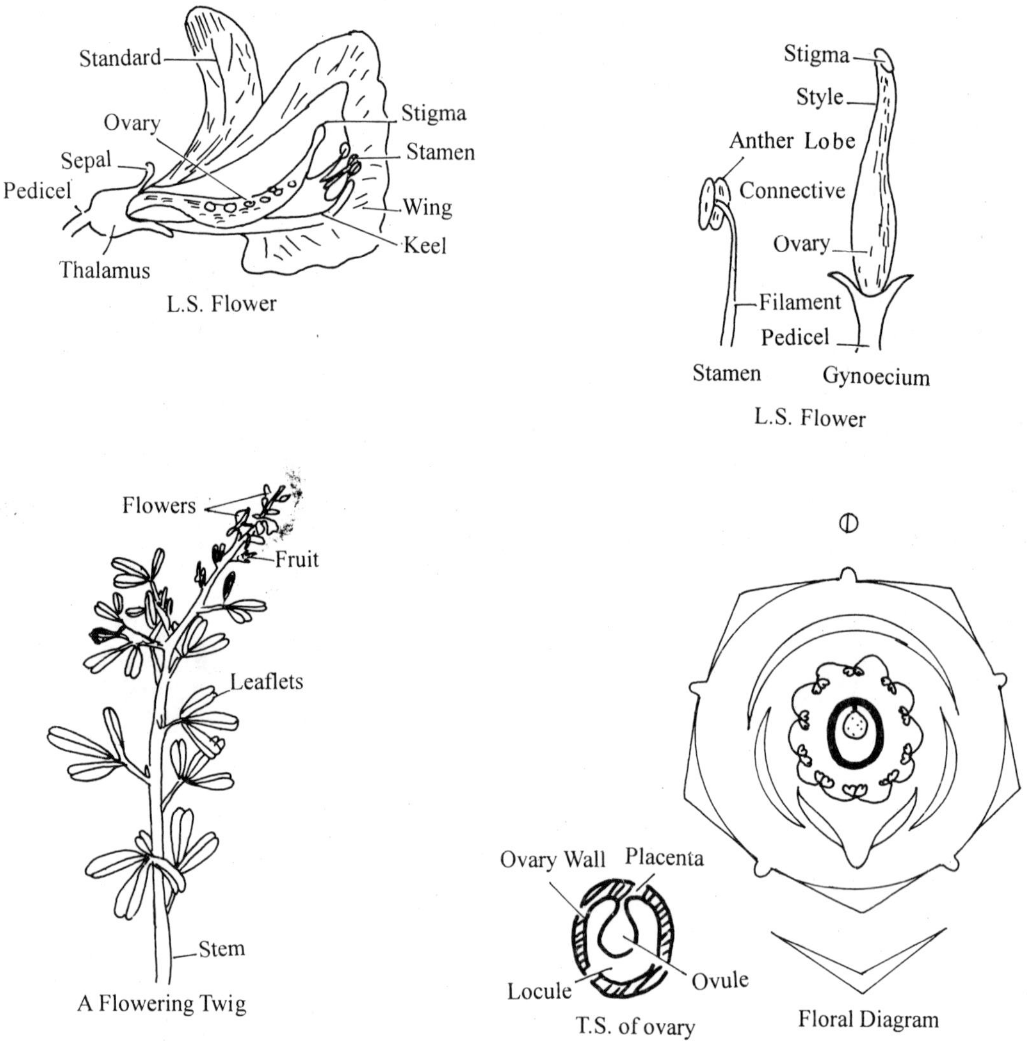

Fig. 11.3 Papilionaceae– *Type Crotalaria Spp*

Leaf : Alternate, stipulate, stipules free, lateral, trifoliately compound and petiolate with pulvinate leaf base; venation multicostate, reticulate.

Inflorescence : Simple raceme.

Flower : Pedicellate, bracteate, bracteolate, complete, heterochlomydeous, zygomorhpic, bisexual and hypogynous.

Calyx : Sepals 5, gamosepalous; odd sepal anterior; aestivation valvate.

Corolla : Petals 5, papilionaceous consisting of a large standard (vexillum) petal, two wing petals (algae) and a boat shaped keel petal made up of the union of two petals. Aestivation descendingly imbricate (standard outermost, keel petal innermost; wings in and out).

Androecium : Stamens 10 in number, monadelphous; anthers dithecous and introrse.

Gynoecium : Ovary superior, monocarpellary and unilocular with many ovules on marginal placenta.

Fruit : A legume.

Floral formula : $K_{(5)} C_5 A_{(10)} G_1$

Family : CAESALPINACEAE (Eg. *Caesalpinia pulcherrima*)

Habit : Cultivated perennial shrub.

Root : Tap root system.

Stem : Aerial, erect, woody, cylindrical, branched solid, glabrous and green in colour.

Leaf : Ramal and cauline, alternate, petiolate, bipinnately compound; paripinnate and stipulate. Leaflets small, eliptical; venation reticulate and unicostate.

Inflorescence : Corymbose raceme.

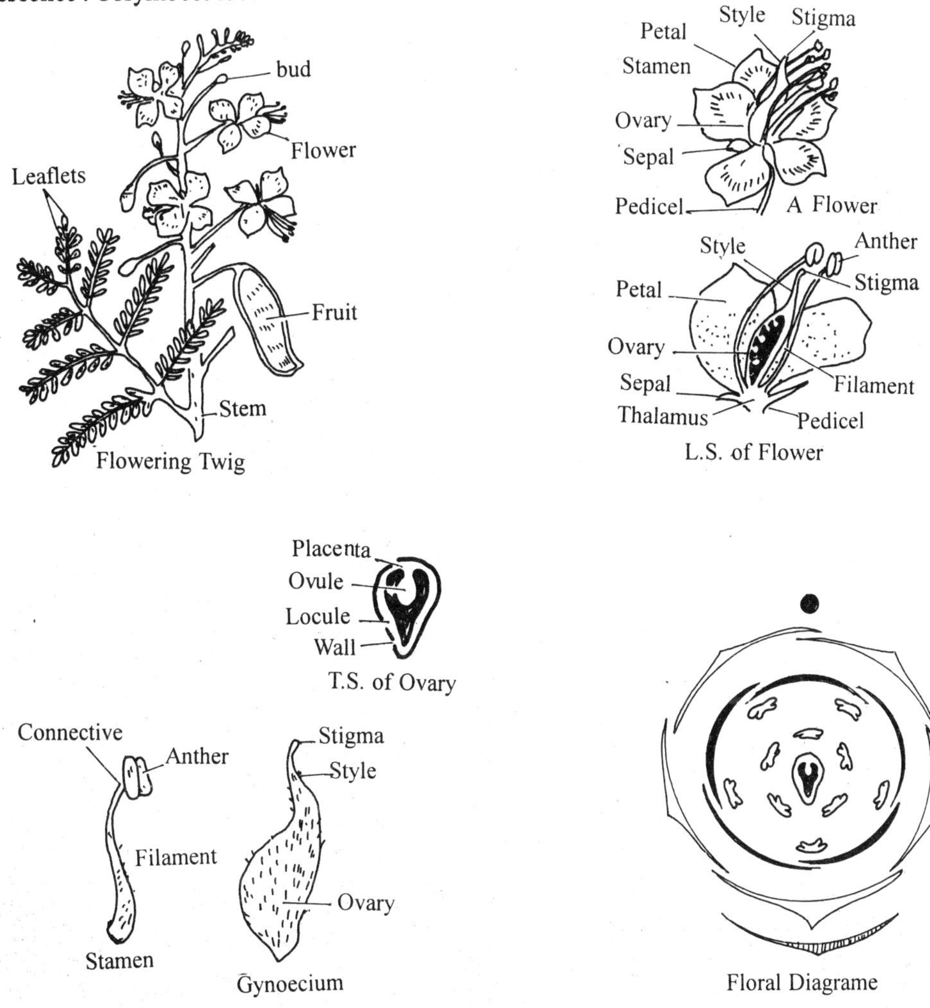

Fig. 11.4 Caesalpinaceae–Type *Caesalpinia pulcherrima*

Flower : Pedicellate, bracteate, ebracetolate, heterochlamydeous, pentamerous, complete, cyclic, zygomorphic, bisexual and hypogynous.

Clayx : Sepals 5 in number, polysepalous, sometimes fused at the base; aestivation imbricate, posterior sepal large and hooded. Corola : Petals 5, polypetalous, odd posterior petal smallest, aestivation ascendingly imbricate.

Androecium : Stamens 10, free, arranged in two whorls of five each, filaments long, anthers dithecous and introrse.

Gynoecium : Ovary superior, monocarpellary, unilocular with many ovules on marginal placenta.

Fruit : A legume.

Floral formula : $K_5 C_5 A_{10} G_1$

Family : MIMOCEAE (Eg. *Mimosa pudica*).

Habit : Small wild perennial herb.

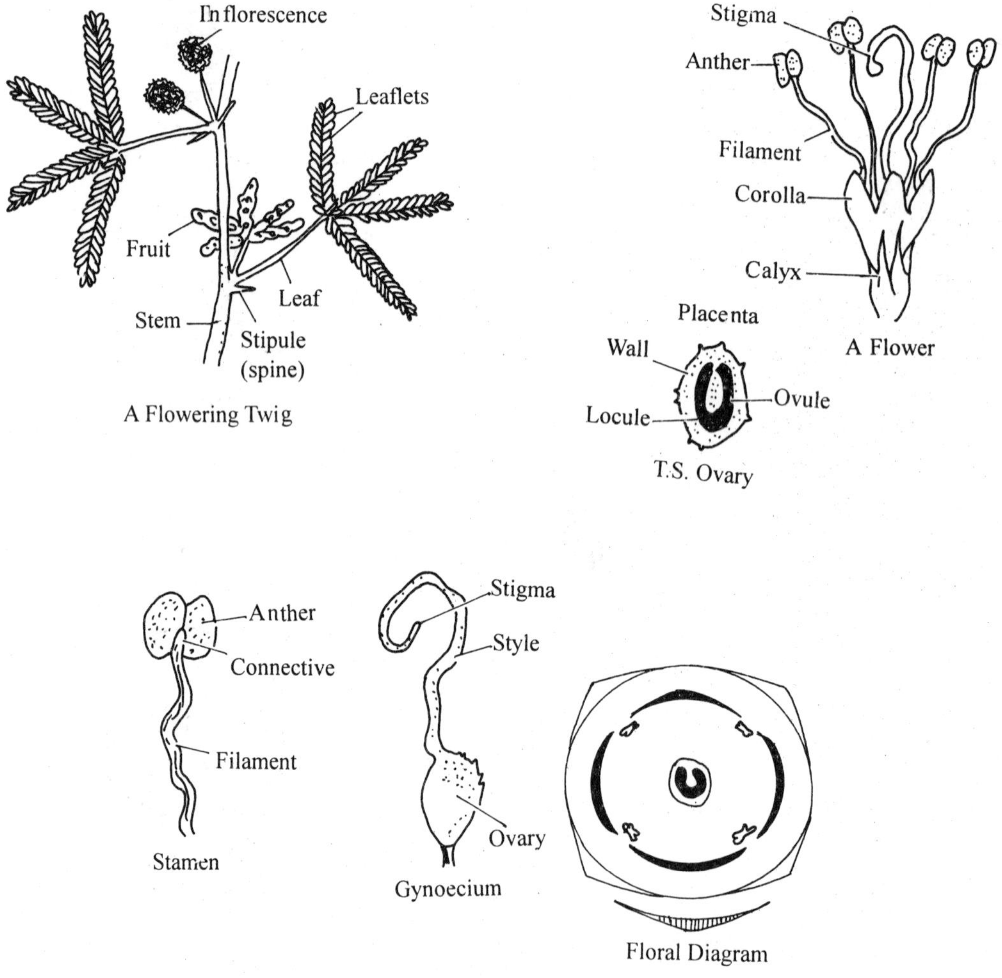

Fig. 11.5 Mimoceae–Type *Mimosa pudica*

Root : Tap root system.

Stem : Aerial, erect, soft, greenish, cylindrical and hairy with epidermal prickles.

Leaf : Petiolate, alternate with a pulvinate base. Leaves are bipinnately compound and paripinnate (leaflets even in number). The leaflets of Mimosa pudica exhibit a kind of movement and drop down on touch, hence the plant is known as 'touch medium not' plant.

Inflorescence : Globose head.

Flower : Sessile, bracteate, ebracteolate, heterochlamydeous, actinomorphic, tetramerous, bisexual and hypogynous.

Calyx : Sepals 4, gamosepalous; aestivation valvate.

Corolla : Petals 4, polypetalous, aestivation valvate.

Androecium : Stamens 4, free; anthers dithecous.

Gynoecium : Ovary superior, monocarpellary, unilocular with many ovules on marginal placenta.

Fruit : A lamentum (a deeply lobed legume).

Floral formula : $L_{(4)} G_4 G_1$

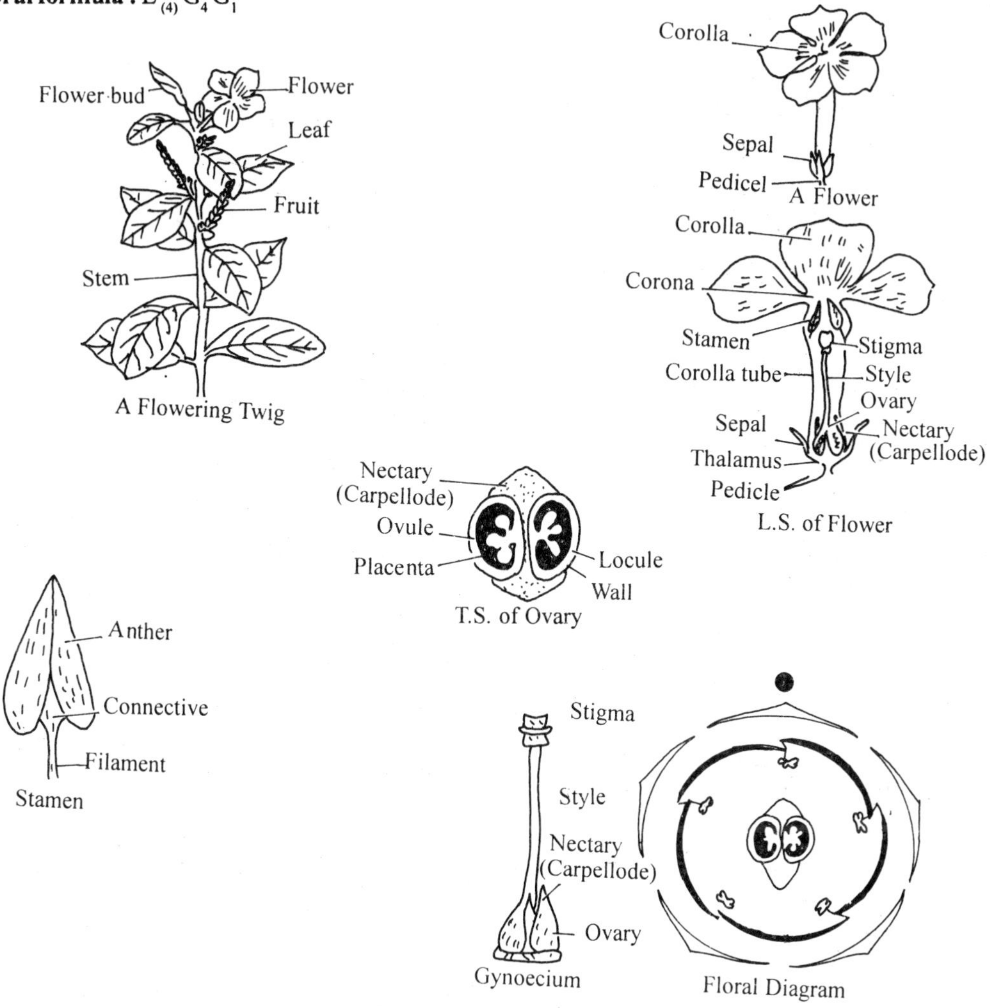

Fig. 11.6 Apocynaceae–Type *Vinca rosea*

Family : APOCYNACEAE (Eg. *Vinca rosea*).

Habit : A perennial cultivated herb.

Root : Tap root system.

Stem : Herbaceous, aerial, erect, angular solid, green and branched. Milky latex present.

Leaf : Ramal and cauline, simple, opposite, decussate and stipulate with inter petiolar stipules. Shape of the leaf obovate, margin entire, venation unicostate reticulate.

Flower : Pedicellate, ebracteate, ebracteolate, complete heterochlamydeous, cyclic, pentamerous, actinomorphic, bisexual and hypogynous. Flowers either pink or white in colour.

Calyx : Sepals 5, polysepalous (fused at the base), persistent and green; aestivation valvate.

Corolla : Petals 5, gamopetalous and hairy corona present at the mout of corolla tube; aestivation twisted.

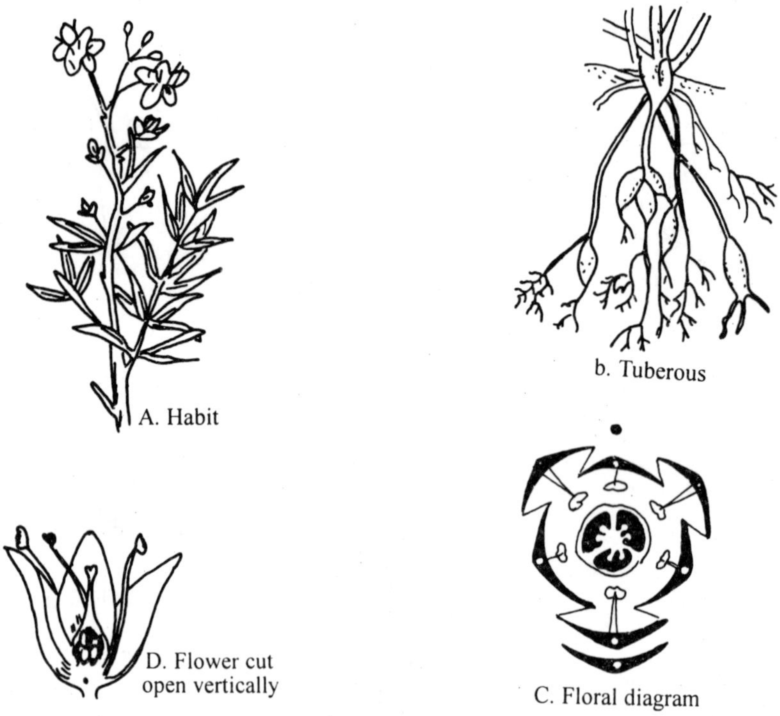

Fig. 11.7 Liliaceae–Type *Asparagus rottlri*

Androecium : Stamens 5, free, epipetalous and inserted at the mouth of corolla tube, filaments short; anthers dithecous and introrse.

Gynoecium : Ovary superior bicarpellary, syncarpous (ovaries are free but carpels are united by the style) and unilocular; placentation marginal. Honey secreting disc present.

Fruit : An etaerio of follicles.

Floral formula : $K_5 C_5 A_5 G_{(1)}$

Family : LILACEAE (eg *Asparagus rotteeleri*)

Habit : A xerophytic climbing herb.

Root : Adventitious root system; roots are modified to store food. They are called 'fasciculated roots'.

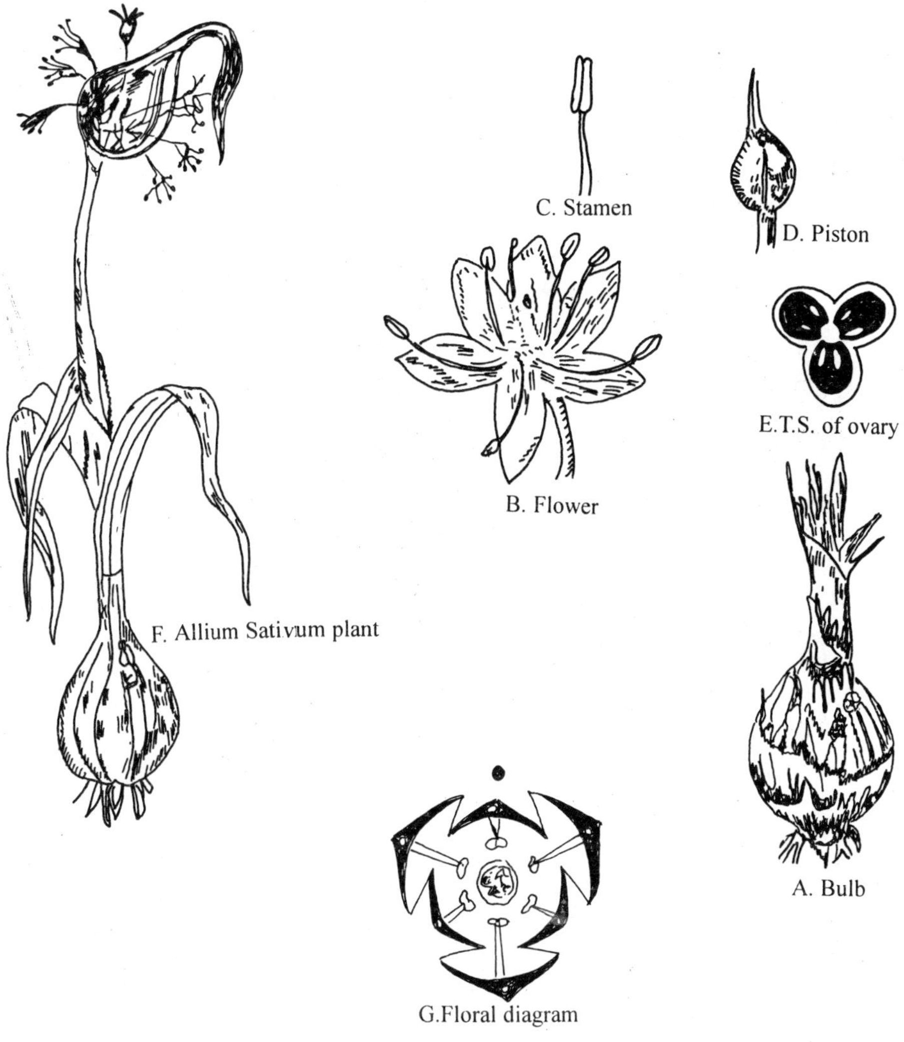

Fig. 11.8 Liliaceae–Type *Allium cepa*

Stem : Aerial, erect, sometimes twining and green in colour; branched, branches of one internodal length clustered at the node forming the cladodes. These are green in colour, look like leaves and help in photosynthesis.

Leaf : Reduced to small scale-like structures.

Flower : Pedicellate, bracteate, ebracteolate, homochlamydeous, complete, actinomorphic, bisexual and hypogynous.

Perianth lobes : Tepals 6, arranged in two whorls of 3 each and free; aestivation valvate.

Andorecium : Stamens 6, free, arranged in two whorls of 3 each and epitepalous. Anthers dithecous.

Gynoecium : Ovary superior, tricarpellary, syncarpous, trilocular with many ovules arranged on axile placenta. Style slender, stigma bilobed.

Fruit : A berry.

Floral formula : $P_{3+3} A_{3+3} G_{(3)}$.

Family : LILIACEAE (eg. *Allium cepa*).

Habit : A cultivated herb.

Root : Adventitious root system.

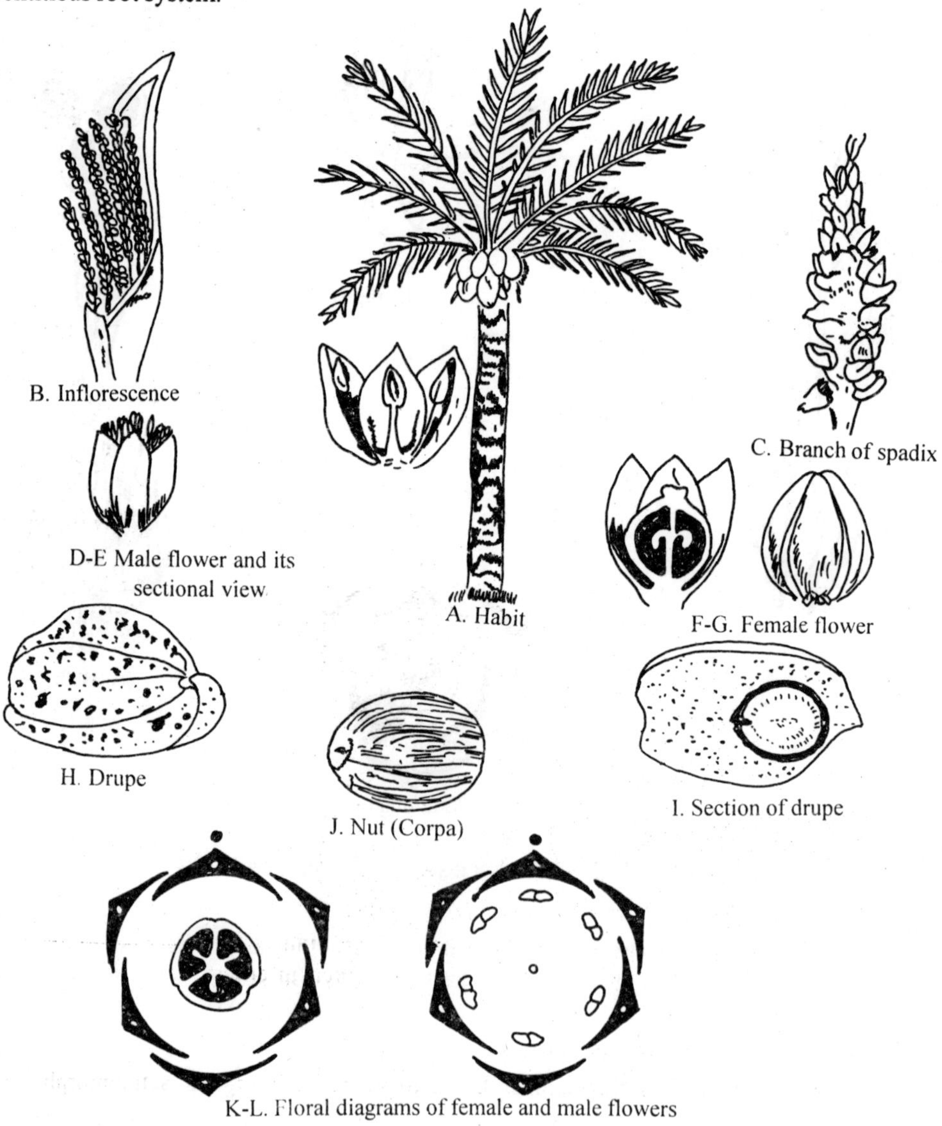

Fig. 11.9 Araceae–Type *Cocos nucifera*

Stem : An underground modification called bulb. Stem reduced to a disc surrounded by fleshy storage leaves.

Leaf : Radical, cylindrical and sheathing. They are two types of leaves, green cylindrical photosynthetic leaves and fleshy storage leaves.

Inflorescence : Looks like an umbel, but it is a collection of several monochasial cymes arranged in an umbellate fashion and enclosed by2-3 membranous bracts.

Flower : Pedicellate, bracteate, ebracteolate, homochlamydeous, trimerous, cyclic, actinomorphic, bisexual and hypogynous.

Perianth : Tepals 6, in two whors of 3 each, white in colour and free; aestivation valvate.

Androecium : Stamens 6, free arranged in two whorls of three each; anthers dithecous.

Gynoecium : Ovary superior, tricarpellary syncarpous, trilocular with two ovules in each locule on axile placenta.

Fruit : A membranous capsule.

Floral formula : $P_{3+3}\ A_{3+3}\ G_{(3)}$

Family : ARECACEAE (PALMAE) eg. *Cocos nucifera* (coconut)

Habit : Tall graceful trees bearing a crown of leaves at the top.

Root : Adventitious root system.

Stem : Tall, erect, woody and cylindrical. The stem is unbranched. It has scars of the leaves.

Leaf : Leaves are fairly large, more than 2 to 3 meters long; unipinnately compound; venation parallel. Lamina is leathery with a hard mid rib. The leaf has a stout petiole with a broad base. When the leaf falls off, a prominent scar is left on the stem.

Inflorescence : A compound spadix: The inflorescence is enclosed by a woody bract (spathe). Each branch of the spadix has a number of male flowers at the top and 3-4 female flowers at the base.

Flower : They are unisexual.

Male flower : Sessile, trimerous, actinomorphic, incomplete, homochlamydeous, unisexual and male.

Perianth : 6 in number arranged in two whorls of 3 each. The perianth lobes ar tough and leathery occasionally fleshy. Outer whorl is smaller than the inner Aestivation valvate or imbricate.

Androecium : Stamens 6, free and arranged in two whorls of three each.

Gynoecium : Absent.

Fruit : Absent.

Floral formula : $P_{3+3}A_{3+3}G_0$

Female flowers : Sessile, actinomorphic, trimerous, incomplete, unisexual fertilization and hypogynous.

Perianth : Same as in male flowers.

Androecium : Absent.

Gynoecium : Ovary superior, tricarpellary, syncarpous, trilocular with one ovule in ecah locule on axile placenta. Usually two of the locules abort and only one develops into the fruit.

Fruit : A drupe.

Floral formula : $P_{3+3}A_0G_{(3)}$

CHAPTER 12

PLANT ANATOMY I

DEMONSTRATION OF MICRO PREPARATIONS BY TAKING TRANSVERSE SECTION OF A DICOT STEM AND LEARNING OF MOUNTING TECHNIQUES

Laboratory studies involve different types of techniques depending on the purpose. If external characters are to be studied of a microscopic organism it is observed under the microscope. If on the other hand internal structure of root stem or leaf has to be studied they have to be cut in order to study the details.

Study of internal structure of a plant organ involves three steps - Section cutting, Staining and Mounting.

Section Cutting: The following is the procedure for taking thin sections for microscopic observation.

Planes of cutting: In order to study the internal structure from various angles sections are cut in various planes.

Transverse: The section is cut at right angles to the longitudinal axis.

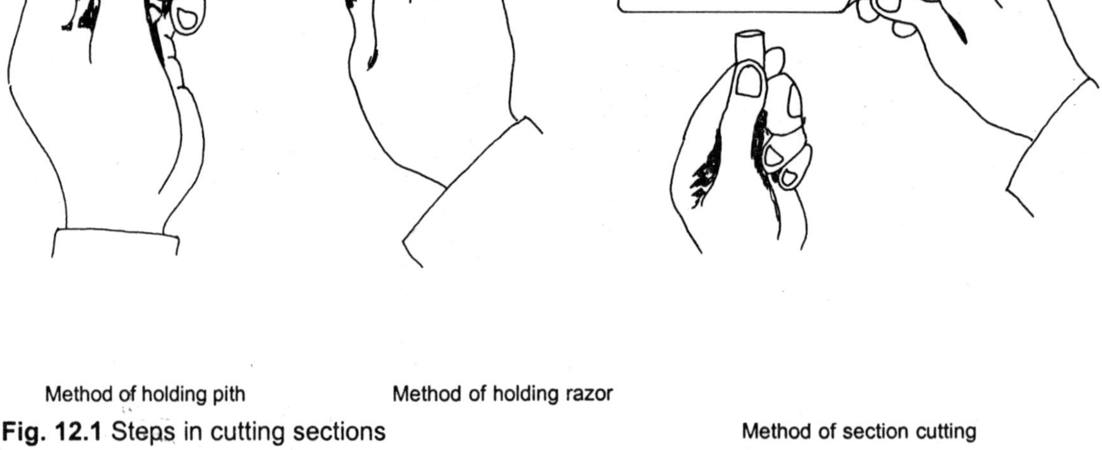

Method of holding pith Method of holding razor

Fig. 12.1 Steps in cutting sections

Method of section cutting

Longitudinal: The section is cut parallel to the long axis(right angles to the transverse axis). The longitudinal section may be cut along a radius (radial longitudinal section - RLS) or along a tangent (tangential longitudinal section - TLS).

Cylindrical organs like roots and stems and dorsiventrally flattened structures(leaves) are cut transversely in order to study the internal details.

Types of section cutting: Usually two types of section cutting methods are employed for plant parts. These are a) free hand sectioning and b) microtome sectioning. Here we will study the details of free hand sectioning. For the purpose of laboratory study where the slides are not made permanent usually free hand sectioning method is employed. Free hand sections are cut mostly with the help of a good sharp razor (sometimes safety razor blades also may be used).

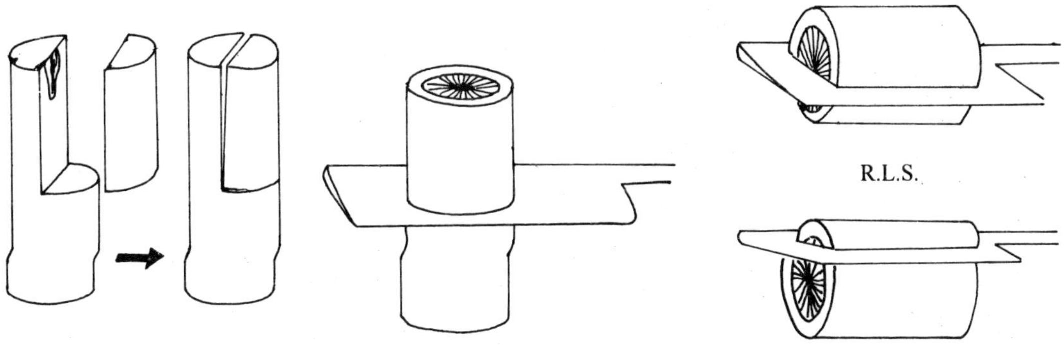

Fig. 12.2 Planes of section cutting

Honing the razor: The razor used for sectioning must have a sharp, even edge. There should be no nicks. Before cutting sections the razor edge has to be honed to maximum sharpness either on a stone or on a leather strap. Oblique strokes with the edge forward are to be given on the stone or the strap. Usually oil drops are used on the honing medium to give maximum sharpness.

Method of cutting: Open out the razor and bend the handle backwards. Hold the razor between the index finger and the thumb. The handle should be free. The remaining fingers of the hand will be touching the back edge of the razor blade.

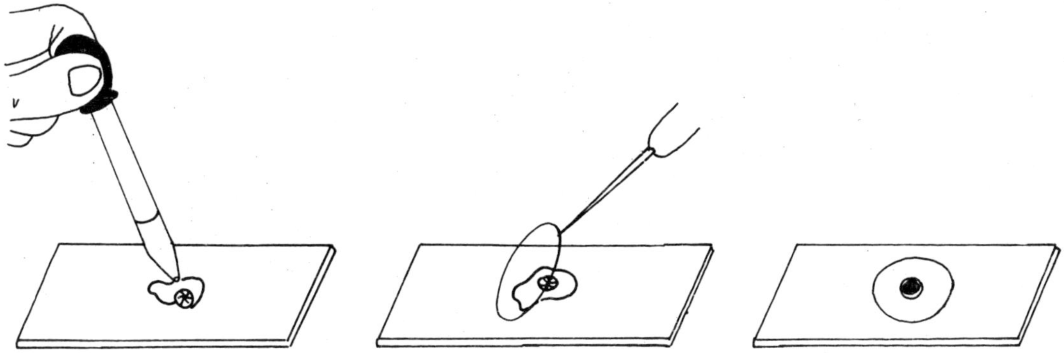

In order to cut a transverse section of a cylindrical part like stem or root follow the procedure given below.

The stem or root should be trimmed into a two or three inches long piece. It should be held in the hand in an erect position between the thumb and the index finger. The material held in the hand should be exactly at right angles to the razor's edge. The thumb holding the specimen should be slightly at a lower level. Flood the edge of a razor with water. Now using the index finger as the platform move the razor quickly several strokes across the material. During this process sections are cut and they will be collected on the blade of the razor. The sections now may be carefully transferred with the help of a fine brush on to a petri dish containing water.

For cutting sections of delicate and fragile materials like the thalli of bryophytes or leaves of angiosperms, a suitable supporting medium has to be used. For this purpose a piece of pith, carrot or potato may be used. These are to be cut into rectangular cubes and split in the centre. The material for sectioning should be inserted into the split and sections are cut in the usual manner using a razor.

After section cutting the razor should be wiped dry. It should be greased and encased.

(In some instances instead of a razor, a shaving blade may also be used but one has to be careful while using the blade as it does not have a handle)

Staining procedure for free hand sections: Free hand sections of root, stem etc. should be washed thoroughly in distilled water in a watch glass. After repeated washing the sections should be transferred to a watch glass containing the main stain. The sections should be kept in the stain for a period ranging from 10 - 30 minutes. After this period the sections should be transferred to a watch glass containing distilled water. The sections should be washed for about 2-3 minutes whereupon excess stain is removed. The sections should be observed under the microscope. If the main stain is saffranin only lignified tissues should retain the stain, others should be destained. If the main stain used is specific to non lignified tissues only they should retain the stain. If the sections are overstained i.e.tissues other than the ones required are still retaining the stain, the sections should be washed thoroughly in water and repeatedly observed under the microscope. The sections can be mildly stroked with a fine hair brush, whereupon excess stain comes out into the distilled water. The sections are now ready for counter staining. They should be transferred to a watch glass containing the counter stain (Fast green etc). Counterstain acts quickly on the tissues. Hence one to two minutes of staining is sufficient. The sections are now observed under the microscope. If there is excess of counter stain it can be removed by washing the sections with 15 - 20% glycerine. After the final staining, the tissues should be demaracted clearly, with no mixing of the stains. The sections are now ready for mounting.

Mounting method: The sections (free hand) are to be mounted on to a clean slide in a specific mounting medium. The sections are to be covered with a thin coverslip or cover glass to facilitate microscopic observation. Cover glasses are readily available in the market; they may be round, square or rectangular. Select the one that is suitable for your preparation.

The sections or objects should be mounted on the centre of the slide. In order to ensure placing of the object on the centre of the slide the following procedure may be adopted on a white paper draw a rectangle of the same size as the slide. Join the opposite corners by drawing diagonal lines. The centre of the slides lies where the two lines cross. Place the slide on this rectangle and the material or object may be kept on the central point.

Place a drop or two of the mounting medium. Hold the coverslip at the edges (see the figure) and allow it to touch the edge of the mounting medium at an angle of approximately 45° C. Slowly lower the cover glass on to the mounting medium with the help of a needle. Take care to see that no air bubbles enter the mounting medium. If there are air bubbles, lift the cover glass and mount again.

Excess mounting medium may be carefully wiped with the help of a clean tissue. Allow the slides to dry. If necessary the slides may be transferred to a hot air oven for drying. The slides should be labelled at one end as shown in the figure. They are now ready for microscopic observation.

Ringing of slides:

Temporary preparations may be kept for few days if the borders are sealed. This procedure is known as ringing. Ringing is done by applying wax, canada balsam or even nail polish to the edges. Ringing may be done manually or by using a ringing table having a rotary metallic disc.

The sections are now ready for microscopic observation.

STUDY OF PLANT TISSUES: Examination of permanent slides of meristem, parenchyma(all types), Collenchyma and sclerenchyma including sclereids.

TISSUES

Cells are grouped together to form tissues. Generally cells which are alike in origin, form and function constitute a tissue. In higher plants like, gymnosperms and angiosperms and to a certain extent in pteridophytes cells are of diverse types. They vary in their origin, shape, size, structure etc., Consequently the assemblage of these different types of cells form various types of tissues.

Basically tissues are classified into meristematic and permanent tissues. The basis for this classification is the stage of development of the cells. In a meristematic tissue the cells are actively dividing and have not attained maturity. The cells of a permanent tissue on the other hand have lost the capacity of cell division and have attained their mature form.

MERISTEMATIC TISSUES

The cells of a meristematic tissue are immature and are in an active state of division. The cells are usually closely packed with no intercellular spaces. The shape of the meristematic cells varies. They may be oval, polygonal rounded or rectangular in shape. The cell inclusions will be in their precursor stage. The nuclei are very prominent. The plane of division in the cells varies. It may be Periclinal (parallel to the surface) or anticlinal (Perpendicular to the surface).

Classification of meristems: There have been different systems of classification based on different criteria, such as origin, location, function etc. A few important types of classifications are given below:

(a) *Meristems based on stage of method of development*: Based on their stage of development meristems may be classified into promeristems and marginal meristems.

Promeristems constitute the region of new growth in a plant body, where the foundation of new organ is initiated. Promeristem is also known as embryonic meristem, primordial meristem or Urmeristem. Structurally the promeristem region consists of initials and their derivatives. The cells have thin cell walls, prominent nucleus, dense vacuolate or non vacuolate cytoplasm. Pits are in their early stages of development. When the cells of the promeristem begin to differentiate they may be said to have lost the characters of a promeristem.

Marginal meristems generally give rise to tissues of cortex, epidermis etc.

(b) *Meristems based on their origin:* Based on the history of the initiating cells meristems are classified into primary and secondary meristems. Primary meristems make up the fundamental parts. They consist of a part of the promeristem. Mainly primary meristems are located at the apices of plant organs.

Secondary meristems always arise late in the ontogeny of the plant. They take their origin in permanent tissues. In position, they are always lateral. Sometimes permanent tissues acquire the capacity of cell division. They are also known as secondary meristems. Typical example of a secondary meristem is the cork cambium.

(c) Meristems based on their location in the plant body: According to the location, there are three types of meristems, viz., apical, intercalary and lateral.

Apical meristems as the very name suggests are located at the apices of shoot, root and other organs. They constitute the growing points of plants. The growth in length of the plant and plant organs is due to apical meristems. Growth is initiated by one or more cells of the apical meristems. Apical meristem consists of two zones, promeristem and meristem. Promeristem consists of actively dividing cells called apical initials. The meristematic zone is classified into protoderm, procambium and ground meristems.

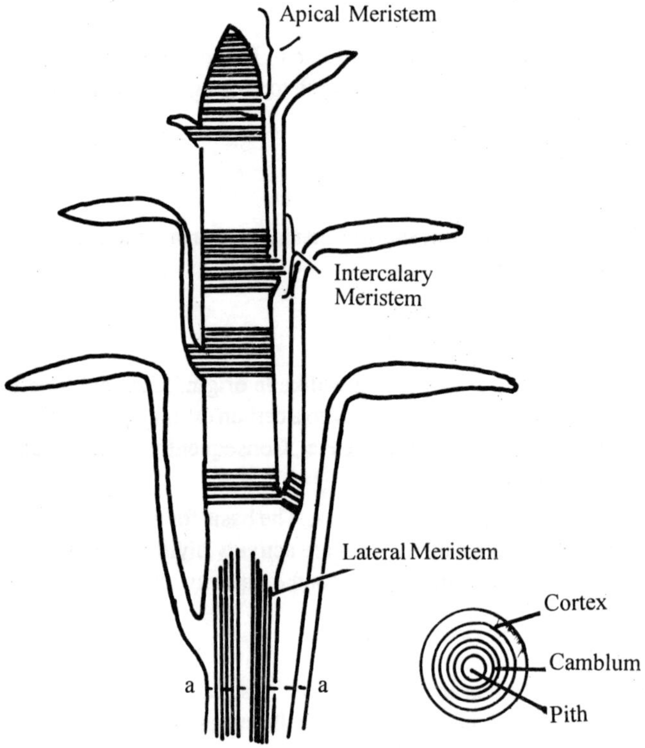

Fig. 12.4 Diagrammatic L.S. (A) and T.S. (B) Showing positions

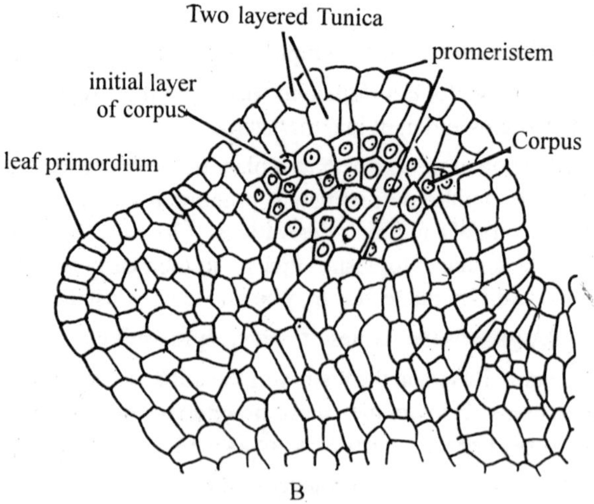

Fig. 12.5 L.S. of shoot apex

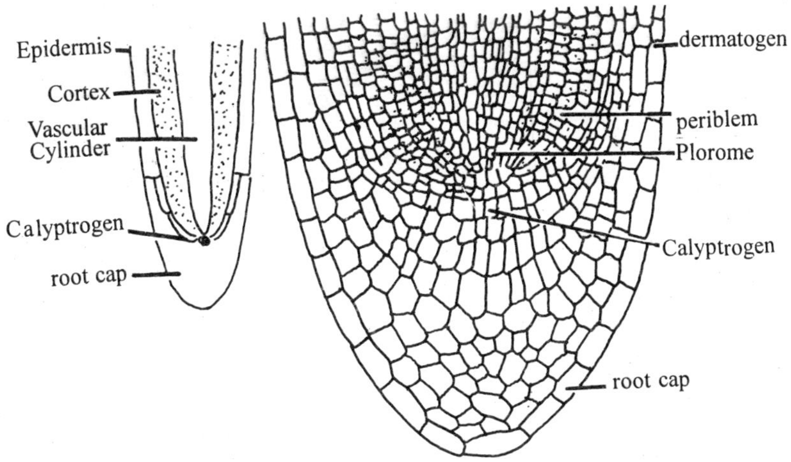

Fig. 12.6 L.S. of root apex

Intercalary meristems represent the detached portions of the apical meristems seperated due to the growth of an organ. Intercalary meristems are always internodal. In the early stages of development of the plant, whole or part of the internode is meristematic. Subsequently a portion of it matures, as a result, patches of meristems are found embedded in permanent tissues.

Lateral meristems are present along the side of the stem. The cells divide in only one plane and add to the girth of the tissues. Typical example of a lateral meristem is the vascular cambium. Vascular cambium produces secondary Xylem and Phloem thus bringing about secondary growth.

(*d*) *Meristems based on their function:* According to the function performed by the derivatives of the meristems, they may be classified into protoderm, procambium and ground meristem. This classification was proposed by Haberlandt at the end of the 19th century. Protoderm develops into epidermis. Procambium gives rise to the primary vascular tissues and the ground meristem are thin and isodiametric. They give rise to hypodermis, endodermis and pericycle.

PERMANENT TISSUES

As a contrast to the meristematic tissues, the cells of permanent tissues have no capacity of division. They have attained their mature form and shape. Very rarely (injury etc.) permanent tissues may become meristematic. This is called dedifferentiation.

The cells of the permanent tissues may be living or dead, thin or thick walled. Permanent tissues may be classified into simple and complex, based on their composition. A simple tissue is homogenous i.e., it is made up of only one type of cells. In a complex tissue there will be at least two different types of cells. But these cells coordinate and function like a single unit. Generally one or more types of simple tissues combine together to form a complex tissues.

SIMPLE PERMANENT TISSUES:

These are further classified into Parenchyma, Collenchyma and Sclerenchyma.

PARENCHYMA: Parenchyma refers to a tissue composed of living cells variable in their morphology and physiology. The word parenchyma is derived from the Greek *para*, beside and *en-chein*, to pour, meaning a semi-liquid substance poured beside other tissues.

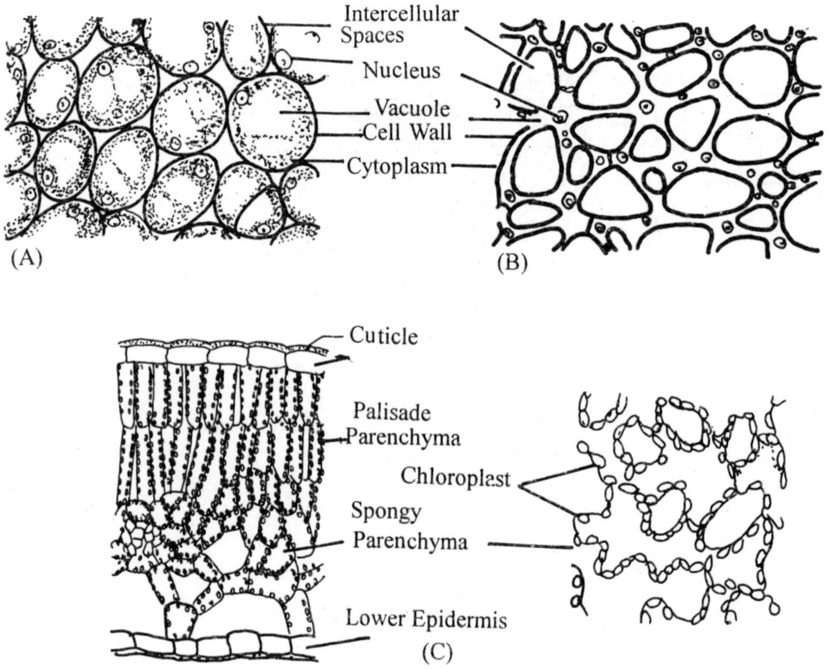

Fig. 12.7 Types of parenchyma

Parenchyma is often considered as the fundamental or ground tissue in which other tissues, notably the vascular are embedded. Ontogenetically parenchyma is the precursor of other tissues. Phylogenetically also it is more ancient than other tissues as seen by the fact that the body of primitive organisms is always parenchymatous. Parenchyma tissue is the principle seat of such essential activities of the plant like photosynthesis, assimilation, respiration, storage, secretion etc.

Developmentally, parenchyma cells are relatively undifferentiated and are unspecialised morphologically and physiologically when compared with other complex tissues.

Distribution: Parenchyma cells are present in almost all parts of the plant, like the pith and cortex of roots and stems, mesophyll of leaves, the flesh of succulent roots, the endosperm of seeds etc. As components of heterogeneous tissues parenchyma cells form vascular rays.

Cell shape: Parenchyma cells are spherical or polygonal in shape (Fig 19). Sometimes they may be elongated and columnar as in the leaves. Studies based on geometric models have indicated that closely packed parenchyma cells have the shape of a polyhedron with approximately fourteen faces.

Parenchyma cells are closely packed or arranged loosely with large number of intercellular spaces.

Cell structure: The cells have thin walls and have dense cytoplasm with a prominent nucleus. Parenchyma cells in storage organs have abundant reserve food materials like proteins, fats and carbohyderates. Parenchyma cells found in leaves have number of choloroplasts . These help in photosynthesis.

Cell types: Based on their structure and function parenchyma cells are classified into the following types

Simple parenchyma which forms the ground tissue, Chlorenchyma which is the chief photosynthetic tissue in the plants, and Aerenchyma which has parenchyma cells with large intercellular spaces storing air. Aerenchyma is present in hydrophytes and helps in respiration and buoyancy.

Chlorenchyma cells found in dorsiventral leaves are of two types viz. Palisade parenchyma and spongy parenchyma. Cells of palisade parenchyma are columnar and are filled with chloroplastss. Spongy parenchyma cells also have chloroplastss, but they are rich in starch grains.

Parenchyma cells found in Xerophytes (*Aloe, Agave* etc.) act as water storage tissues.

COLLENCHYMA: Collenchyma is a living tissue and is composed of more or less thick primary non lignified walls. Just as in parenchyma, even in collenchyma the protoplast is living. The word collenchyma is derived from the Greek word *colla*, meaning glue, which refers to the thick glistening wall

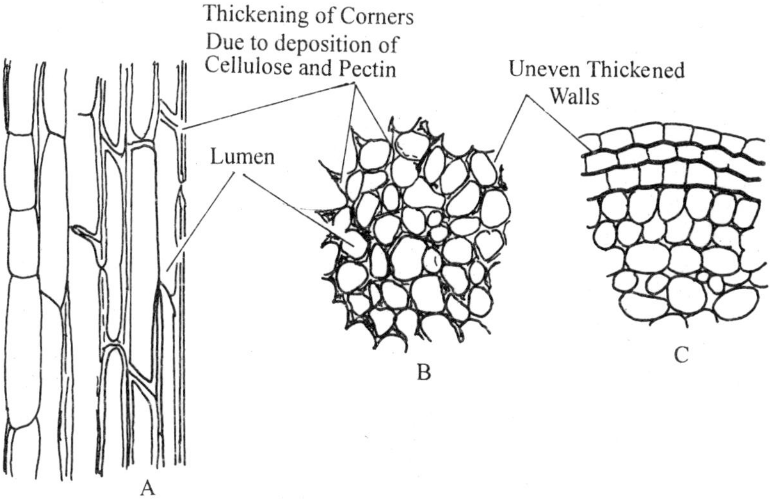

Fig. 12.8 Collenchyma

Distribution: Collenchyma characterstically occurs in the pripheral position of stems, leaves, floral parts etc. It may also occur in root cortex particularly if it is exposed to light. In the stems and leaves of dicotyledons, collenchyma is hypodermal situated immediately below the epidermis or seperated from it by one or two layers of parenchyma. Collenchyma cells are considerably elongated. Sometimes, cells are 2 mm, long resembling fibres. The shorter collenchyma cells are prismatic and appear polygonal in a transection (Fig 1) In a long section the cells appear rectangular.

Cell structure and function: The structure of the cell wall is the most characterstic feature of the collenchyma cells. The thickenings are deposited unevenly, mainly deposited at the corners where several cells join together. In long section, collenchyma cells show thin and thick wall portions dependng on the direction of the cut. The walls containing mainly of cellulosic and pectic compounds and consist of much water. Ultrastructurally the thickenings of collenchyma show an alternation of layers of longitudinally oriented microfibrils and non cellulosic material.

Collenchyma cells have living protoplasts and may have chloroplastss. Chloroplastss are less in number in long and narrow cells and more in cells which approach parenchyma in form. Tannin may be present in collenchyma cells.

Collenchyma is primarily a mechanical tissue adapted for support of growing organs. The thick walls and close packing, make collenchyma a strong tissue. At the same time, the peculiarities of growth and the structure of its walls impart a considerable tensile strength with flexibility and plasticity.

SCLERENCHYMA: This tissue consists of thick walled cells, (lignified), their main function being mechanical. The word is derived from the Greek *Sclerous*, hard and *Enchyma* an infusion which emphasizes the hardness of the cell walls. Haberlandt (1914) and Foster (1949) combine collenchyma and sclerenchyma under a common name stereome which includes the entire mechanical system of the plant.

Sclerenchyma cells show variety in form, structure, origin and development, which the different types often integrating. Generally, sclerenchyma cells are classified into fibres and sclereids. Fibres are considered as long cells while sclereids, are relatively short cells.

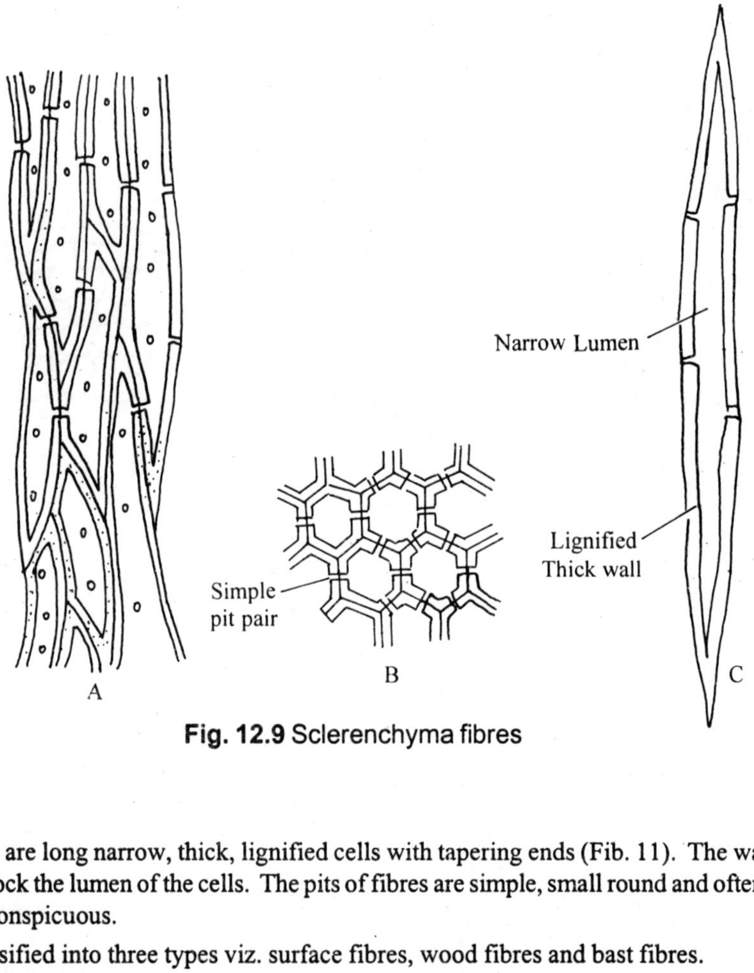

Fig. 12.9 Sclerenchyma fibres

Fibres: Fibres are long narrow, thick, lignified cells with tapering ends (Fib. 11). The wall thickenings sometimes even block the lumen of the cells. The pits of fibres are simple, small round and often oblique. The middle lamella is conspicuous.

Fibres are classified into three types viz. surface fibres, wood fibres and bast fibres.

Surface fibres are found in the seed coats or the pericarp of the fruits (eg cotton fibres, fibres present in the mesocarp of Cocos nucifera).

Wood fibres are present in xylem. These are also called xylary fibres (For details see xylem)

Bast fibres are obtained from the pericycle and phloem of various plants. In *Cannabis sativa* the pericycle yields fibres, which is used in the making of carpets, ropes, sacs etc. Phloem fibres are seen in many plants like *Linum* (for details see phloem)

Fibres are of great economic value. They find a place of pride in textile industry. To the plants, the fibres are of great use as they offer mechanical support.

Besides, xylem and phloem, fibres also occur as bundle caps in *Zea, Saccharum* etc. In *Zea mays* (Murdy 1960) a hypodermis containing fibres as long as 1mm, has been recorded. In palms. Central cylinder is demarcated by a sclerotic zone that may be several cm wide. Fibres may be present in the leaves of monocotyledons, where they form sheaths enclosing the vascular bundles of strands extending between epidermis and vascular bundles.

Sclereids: These are short isodiametric cells (Fig 12) but sometimes may be elongated also. The cell walls are very thick and lignified but the cells do not have tapering ends. The cells are dead and have a very narrow lumen. The secondary walls show prominent pits which are simple. Sometimes there is slight overarching of the secondary wall over the pit chamber. The lignin deposition is uniform, but may be uneven as in the seed coats of some Leguminosae.

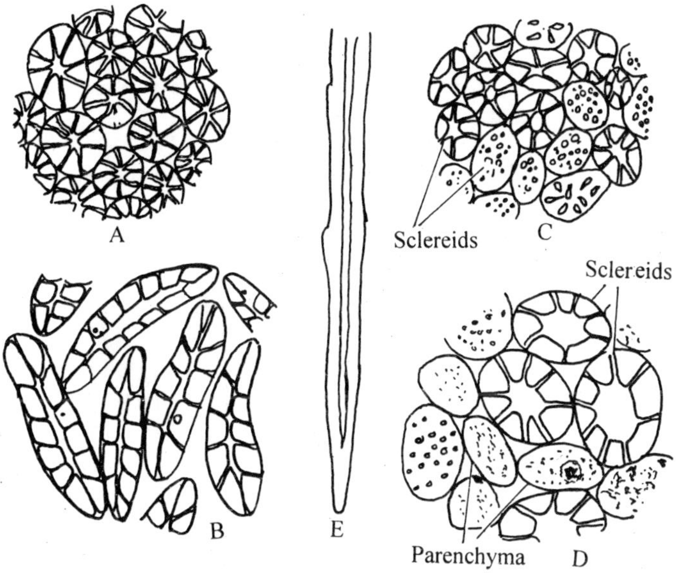

Fig 12.10 Sclereids

Sclereids arise through a belated sclerosis of ordinary parenchyma cells or directly from cells called sclereid primordia.

Sclereids are widely distributed in the plant body. The cortex and the pith of gymnosperms and dicotyledons very often contain sclereids arranged singly or in groups. Xylem and phloem may often contain sclereids besides fibres. Sclereids may be present in leaves, seeds and fruits. In fruits they are sometimes dispersed in the soft tissues.

Sclereids are classified into the following types viz., brachysclereids (fig 12) or stone cells (short isodiametric cells widely distributed in the flesh of fruits), macrosclereids long rod like cells found in the epidermis of leguminous seeds (Fig 12) Osteosclereids (bone like cells) Astrosclereids (star like cells) Bizarre sclereids (Irregular sclereids), Filiform sclereids (long cells resembling fibres and trichosclereids branched, thin walled sclereids).

STUDY OF PLANT TISSUES (Contd.) : *Examination of permanent slides of xylem and phloem, Maceration technique for Xylem tissue and observation of vessels, trachieds and fibres.*

COMPLEX PERMANENT TISSUES

The vascular tissues are the most important among the complex tissues. There are mainly two components in the vascular tissue viz., xylem and phloem.

XYLEM: This is mainly a conducting tissue of water and mineral nutrients. The term xylem was introduced by Nageli (1858) and is derived from the Greek *Xylos* meaning wood. Structurally the xylem is most complex, because it consists of various types of living and non-living cells.

The xylem is mainly classified into primary and secondary xylem. The former differentiates in the primary plant body and is derived from the procambium, while the latter originates from a fascicular cambium during secondary growth.

Elements of xylem: The xylem is composed of different kinds of elements viz., tracheids, fibre tracheids, vessels or tracheae, wood fibres and wood parenchyma.

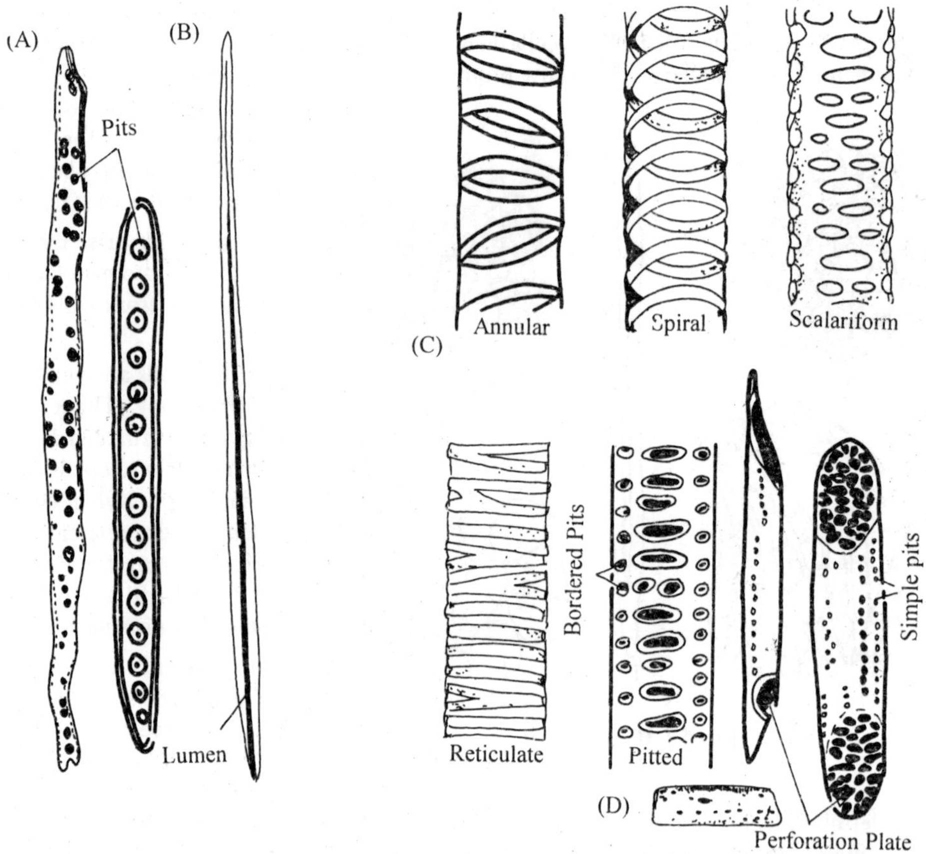

Fig. 12.11 Xylem

(a) Tracheids: They are the fundamental type of xylem element present in pteridophytes and gymnosperms where they constitute the chief conducting elements. Occassionally, they occur in Angiosperms also (Magnoliaceae). A tracheid is an elongated tube like cell (Fig 12.11) with tapering rounded ovel ends. The tip of the tracheid of a secondary xylem may be chisel-like. The walls are thickened due to the deposition of lignin, which are of various types.

Thickening of the cell wall: The secondary wall acquires different types of thickenings. This is specially seen in cells which are to differentiate into tracheary elements i.e, vessels and tracheids. In these cells, the thickening is due to the deposition of lignin. The deposition of lignin is not uniform, hence we find different types of thickenings. Chiefly, the following types of thickening are seen in xylem.

Annular: Lignin deposition is in the form of annuli or rings. These rings are unconnected. The space between the rings is unthickened. (Fig 12.11A)

Spiral: Lignification occurs in the form of a spiral band (Fig 12.11B)

Scalariform (ladder like): Wall thickening is in the form of transverse rods, very much resembling the rungs of a ladder (Fig 12.11C).

Reticulate: Lignin is deposited in the form of network or reticulum. The unthickened areas of the cell wall are irregular in shape (Fig 12.11D)

Pitted: Lignin deposition is almost uniform over the wall, except for small pores here and there. These pores are called pits. (Fig 12.11E)

Structure of Pit: Pits are formed on either side of the wall. In any area there will be two pits lying opposite to each other. These are usually termed as "pit pairs". There is a space inside each pit called pit cavity or pit chamnber. There is a thin membrane which seperates the pit cavities of the opposite pits. It is called pit membrane. The opening to a pit is called the pit aperture.

A pit membrane is common to both the pits. It has a primary wall on either side seperated by middle lamalla.

There are two types of pits viz. simple pits and bordered pits. A simple pit is just an opening in the wall. Whereas in a bordered pit, the pit aperture is overarched by the projection of the secondary wall. In a pit pair both may be simple, both may be bordered or one may be simple and the other one bordered. The lost mentioned one is generally called a half bordered pit pair. A pit that occurs opposite to an intercellular space will not have its complementary and hence known as a blind pit.

Simple pits generally occur in parenchyma cells, medullary rays, phloem fibres, companion cells, and also in tracheids, pit cavity is uniform in its diameter and the pit membrane is simple. Simple pits may vary in shape (oval, Polygonal, irregular etc.)

In a bordered pit, the pit membrane becomes thick at the centre. The thickening is known as the torus. The pit membrane may also change its position. The torus instead of remaining at the centre may shift tyowards a pit aperture. In bordered pits the aperture is narrow (pit canal) and it widens towards the interior (pit chamber). Bordered pit occurs in the vessels of Angiosperms and in the tracheids of many Conifers.

Elements of xylem:

Tracheids: These are dead cells and like sclerenchyma have neither the nucleus nor the protoplast. The lumen of the cell is large. The wall of the tracheids is provided with a number of bordered pits.

In a cross section tracheids appear polygonal or rectangular. Primary xylem tracheids develop from the procambial strands.

Besides helping in conduction, tracheids also give mechanical support to the plants. Tracheids rarely exceed 1 mm in length, though, there are reports of tracheids more than 10mm long (Conifers).

(b) *Fibre tracheids* : During the development of fibres the wall thickening increases whereas the lumen decreases. In many cells, the lumen becomes so narrow that it is altogether obliterated. In such cells there is little or no conduction and they may be said to have reached the stage of true fibre.

(c) *Vessels*: These are evolved from tracheids and are present mainly in Angiosperms. However., there are several instances of their presence in pteridophytes (*Selaginella, Pteris* etc.,) and Gymnosperms. Vessels arise through the dissolution of the end walls of tracheids . Thus sometimes a tracheids may represent a developmental stage of a vessel. Vessels are composed of rows of cells placed one above the other with the transverse walls being eliminated. This facilitates the easy flow of water. Vessels may arise from tracheids either by the dissolution of the pit membranes which makes the end wall preforated or by the dissolution of the entire wall.

Vessels rarely exceed a length of 10cm. Occassionally however they may be two meters long (*Quercus*) or even 3-6 meters long (*Eucalyptus*). Like tracheids vessels also have different types of wall thickenings due to the uneven deposition of lignin. Vessels also have a wide lumen.

Vessels (tracheids also) may be of two types protoxylem and metaxylem. The former matures a little early in the ontogeny and has a narrow lumen while the latter matures later and has a broad lumen.

(d) *Wood fibres*: They are also called xylary fibres. sclerenchyma fibres associated with xylem are termed wood fibres. They possess very thick walls and few small pits. They originate from the derivatives of the cambium and have varied shapes. In wood fibres, known as libriform fibres, the wall is extremely thick and lumen is almost absent. Like the tracheids , wood fibres are also dead and their main function is mechanical support.

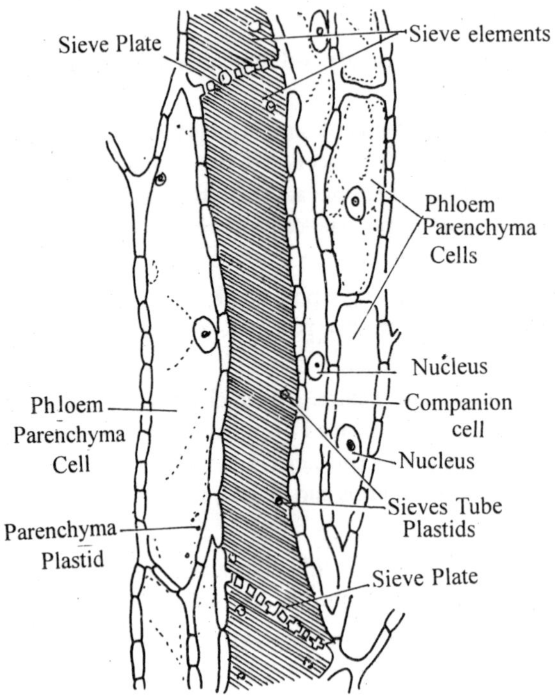

Fig. 12.12 L.S. of phloem tissue

(e) Wood Parenchyma: PARENCHYMA cells assicoated with xylen are called wood parenchyma . These are living cells and remain so, as long as the tissue in which they lie conducts water. The cells of wood parenchyma may be thin or thick walled have plenty of storage food and may help in conduction. In secondary xylem wood parenchyma forms the rays which may help in radial conduction.

PHLOEM : The phloem is the principal food conducting tissue in vascular plants. Unlike xylem all the elements of phloem (except for phloem fibres) are living. Like the xylem phloem is also of two types of viz., primary phloem (derived from procambium) and secondary phloem (derived from fascicular cambium). The basic components of phloem are sieve elements, companion cells, phloem parenchyma and phloem fibres. Occasionally sclereids, laticiferous cells and idioblasts are also found in phloem.

(a) Sieve elements: What the tracheids are to xylem sieve elements are to phloem. Sieve elements may be seggregated into relatively less specialized sieve cells (found in pteridophytes) and more specalized sieve tubes. Seive tubes are living cells and have a thin cellulose wall. In angiosperms they are enucleate, while they are nucleate in gymnosperms.

The protoplast of a sieve tube has a big central vacuole which is filled with viscous and albuminous proteins. Leucoplasts are also reported in the vacuole.

The cross walls between the seieve tubes are porus and are called sieve plates as they resemble a sieve (fig 14). Sieve plates are usually of two types viz., simple and compound. In simple sieve plates, the pores are large and the plate is horizontal or transverse in position. The compound sieve plate is somewhat oblique. The sieve pores are large and the plate is horizontal or transverse in position. The compound sieve plate is somewhat oblique. The sieve pores are very often plugged with a shining substance called callus pad which is usually deposited at the end of the spring season. Chemically callus pads are made up of a carbohydrate called callose. At the approach of the spring, the callus dissolves. Through the sieve pores, thin cytoplasmic strands come out and enter into other cells.

(b) Companion cells: These are specialised small cells usually associated with the sieve tubes. These are living, nucleate cells and are supposed to compensate the absence of nuclei in sieve tubes. Companion

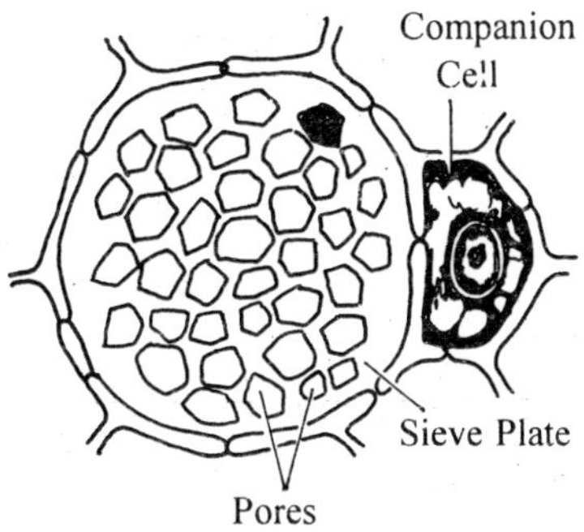

Fig. 12.13 T.S. of seive plate

cells are absent in gymnosperms, where albuminous clls are present. The cytoplasm of cmpanion cell is dense and granular with a prominent nucleus. Ontogenetically both sieve tubes and companion cells are derived from the same mother cell.

(c) Phloem parenchyma : PARENCHYMA cells in association with phloem are called phloem parenchyma. They are no different from other ordinary parenchyma cells.

In shape, the cells may be elongated, pointed, broadly cylindrical or even subspherical. The cells contain oil, starch mucilage, latex etc. Phloem parenchyma is absent in most of the monocotyledons.

(d) Phloem fibres: These are also called bast fibres. These are found both in primary and secondary phloem. The walls have simple pits. The bast fibres may be lignified or non lignified. The bast fibres of Linum usitatissimum are cellulosic and are of great value in commerce. The fibres of Linum are derived from protophloem and are arranged in irregular bands.

In order to observe the structural details of xylem and phloem either a cross section or a long section may be taken. Sometimes in order to study the entire cells without cutting, a special technique called maceration technique is followed. In this technique the plant parts like stem or root pieces are treated with concentric acids wherein the middle lamella will be digested and the cells become free. These cells can then be separated, stained and mounted on a slide for microscopic observation. The following is the general technique of maceration.

Maceration:

1. Take small slices of fresh or dry material of stem, root or leaf.
2. Boil and cool in water repeatedly.
3. Transfer the materials into a tube containing the maceration mixture - equal parts of aqueous nitric acid (10%) and aqueous chromic acid.
4. If the tissue is very hard, heat the solution in a hot air oven at 30 to 40 degree centigrade until the tissue becomes soft.
5. Gently crush the tissue with a thick glass rod having a rounded end.
6. If the cells are not still separated transfer the material to fresh macerating fluid and repeat step 4.
7. Wash thoroughly in water to remove acids. If necessary use a centrifuge.
8. Take out a little of the macerated tissue, spread properly on the slide and stain with safranin.
9. Rinse in water, dehydrate and mount in glycerine. Cover the preparation with a cover glass.
10. Observe under the microscope and make sketches.

CHAPTER 13

PLANT ANATOMY II

STUDY OF PLANT ANATOMY: Examination of Vascular bundles; root anatomy of monocots (Maize) and dicot (Helianthus).

Types of Vascular bundles

A vascular bundle may be defined as grouping together of vascular elements namely xylem and phloem to form a functional unit. Within a vascular bundle the two major components - xylem and phloem are arragned in various ways. Depending on their relative position vascular bundles are categorized into the following types - cambium which brings about secondary growth may or may not be present.

Open vascular bundles : This is seen in dicotyledons. Here between xylem and phloem, cambium (secondary meristem) is present. As a result of this the vascular bundle is open to secondary growth. There is no bundle sheath surrounding the vascular bundle.

Closed vascular bundle : This is seen in monocotyledons. Here there is no cambium in the vascular bundle. Hence there is no secondary growth. Further a bundle sheath surrounds the vascular bundle.

Based on the relative positions of xylem and phloem vascular bundles are classified into- concentric, radial and conjoint types.

Concentric vascular bundle: In concentric vascular bundles xylem and phloem form concentric rings; either xylem surrounds the phloem or phloem surrounds the xylem. Concentric vascular bundles are of two types-

Amphivasal (Leptocentric): Here phloem is in the centre surrounded by xylem ec., *Dracena, Yucca* etc.

Amphicribal (Hadrocentric) : Here xylem is in the centre surrounded by phloem eg. Ferns.

Concentric bundles are always closed ie., they have no cambium.

Radial bundles: This kind of a vascular bundle is usually seen in the roots of dicotyledons and monocotyledons. Here the xylem and phloem together will not form a single bundle, instead they form alternating bundles and are present on different radii. This is regarded as the most primitive type.

Conjoint bundles : In this type xylem and phloem together form compact bundles and they are present on the same radius. Conjoint bundles may be open or closed and are usually present in stems. Conjoint vascular bundles of monocots are always closed (no cambium) while those of dicots are always open (cambium present).

Conjoint vascular bundles are of two types.

Collateral : Here xylem and phloem lie on the same radius with xylem towards the centre and phloem towards the periphery. In otherwords there is one group of phloem and one group of xylem. In dicotyledons cambium is present in between xylem and phloem while it is absent in monocotyledons. Collateral vascular bundles are found in stems of Angiosperms.

Bicollateral : In these bundles there is one group of xylem surrounded by two groups of phloem - outer and inner. Similarly two strips of cambia are also present. Here the arrangement of various elements are as follows- inner phloem, inner cambium, xylem, outer cambium and outer phloem. Bicollateral vascular bundles are usually seen in the members of the family Cucurbitaceae.

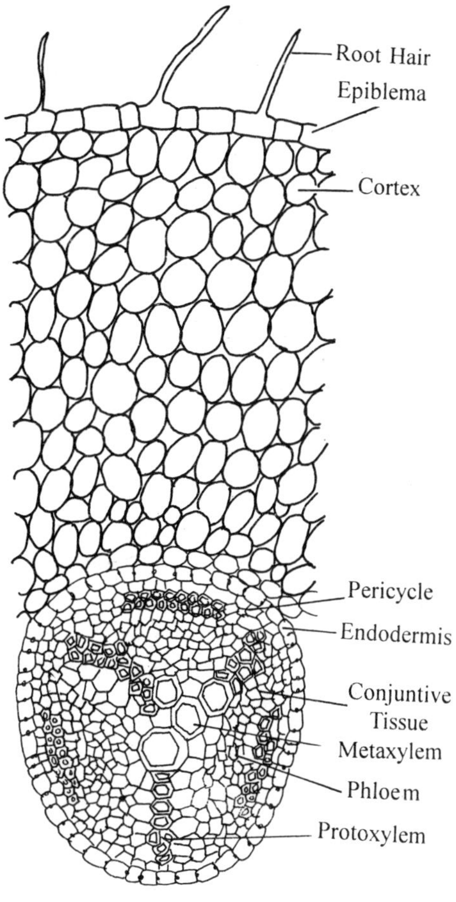

Fig 13.1 T.S. of Dicot root (Helianthus)

INTERNAL STRUCTURE OF ROOTS

Root of Helianthus:

As an example of a typical dictylendonces root, the root of sunflower may be studied.

A transverse section of a young sunflower root shows the following features (Fig 16).

Epidermis: This is the outermost layer. This is also known as epiblema or piliferous layer and is made up of closely packed cells with no intercellular spaces. Some of the epidermal cells produce unicellular extensions called root hairs. These help in the absorption of nutrients from the soil.

Cortex: Internal to the epidermis is the cortex. It is made up of several layers of thin walled spherical or oval parenchyma cells with large number of intercellular spaces. Some of the outer layers of the cortex may become thick walled and function as exodermis.

Stele: Occupying the central region of the root is the vascular cylinder. The outermost layer of the stele is the endodermis. It is made up of a single layer of barrel shaped cells., closely packed without any intercellular spaces. The endodermal cells have thickenings on their radial walls called casparian thickenings. In between endodermal cells thin walled cells are present. They are called the passage cells. Usually they are present opposite the protoxylem.

Internal to the endodermis is the pericycle. It is made up of a single layer of rough, thin walled parenchyma cells.

Surrounded by the pericycle are the vascular bundles. They are radial (Xylem and Phloem form separate alternating bundles) with *exarch* (protoxylem pointing towards the periphery) and *tetrarch* (four protoxylem points) xylem. Xylem consists of annular and sprial vessels with metaxylem being made up of reticulate and pitted vessels.

Occupying the central region of the stele is a narrow parenchymatous pitch. Sometimes the metaxylem groups join together so that pith is obliterated.

Between the radial vascular bundles is present parenchyma constituting the conjuctive tissue.

ROOT OF MAIZE: As an example of a typical monocotyledonous root, the root of maize plant m ıy be taken. A transverse section of a young root shows the following features. Mainly three regions may be distinguished. These are epiderms, cortex and stele.

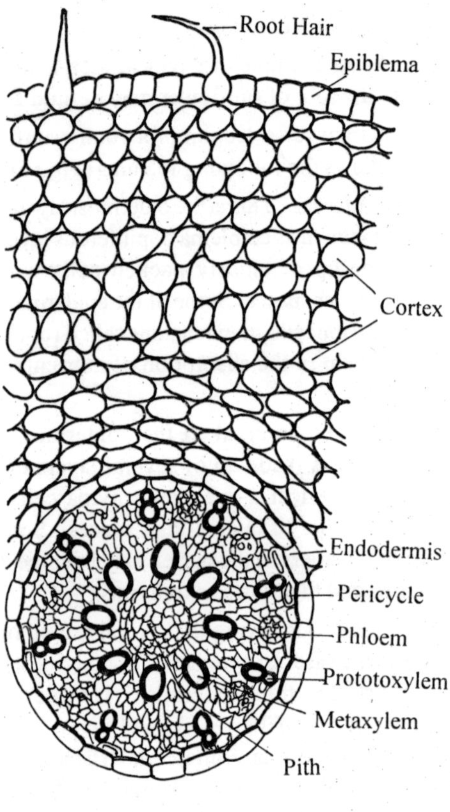

Fig 13.2 T.S. of Monocot (Maize) root

Epidermis: This is the outermost layer, and is also known as epiblema or piliferous layer. The epidermis is made of a single layer of cells which are closely packed without any intercellular spaces.

Cortex: Immediately internal to the epidermis is the cortex which is made up of many layers of loosely arranged parenchyma cells. Some of the cells in the outer region of the cortex may be closely packed constituting the hypodermis.

Stele: The stele is bounded by an endodermis and pericycle. Endodermal cells are barrel shaped and have casparian thickenings on their radical and tangential walls. Pericycle is represented by a single layer of thin walled parenchyma cells.

The vascular bundles are many in number (polyarch) and radial with exarch xylem. In the centre of the stele is a parenchymatous pith.

The main points of difference between the dicotyledonous root and the monocotyledonous root are the number of xylem groups and the presence or absence of pith.

STUDY OF PLANT ANATOMY (contd): Stem anatomy - Dicot (Helianthus) and Monocot (Maize) stems.

ANGIOSPERMS - ANATOMY

Internal structure of stems

Internal structure of plants organs like roots, stems and leaves may be studied by taking either a long section or a transverse section. In a transverse section all the tissues will be cut crosswise and they give us an idea of their structure, distribution etc. In this chapter we will discuss the primary and secondary structure of roots, stems and leaves.

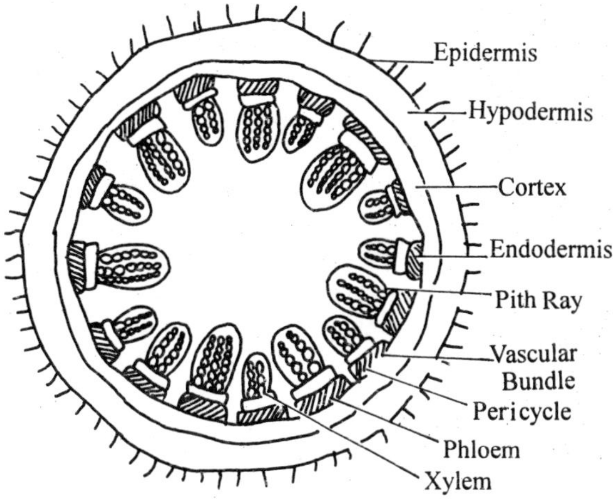

Fig. 13.3 Ground plan of T.S. of dicot stem (Helianthus)

STEM OF HELIANTHUS (Sunflower):

A transverse section of young sunflower stem shows an outmost epidermis, a middle cortex and a central stele (Fig. 18).

Epidermis: This is single layered with closely packed rectangular cells. Above the epidermis there is a well defined cuticle (Fig.19). Some of the epidermal cells produce multicellular hairs. Occasionally stomata may be seen in the epidermis.

Cortex: These are two regions in the cortex. The outer cortex (immediately below the epidermis) consists of a few layers of collenchyma whose corner walls are thickened by the deposition of pectin and cellulose. Next to the collenchyma is parenchyma constituting the inner cortex. This consists of several layers of parenchyma , some of the outer layers of which are chlorophyllous.

Stele: The outermost layer is the endodermis whose cells are rich in starch grains; hence also known as the starch sheath. It has the thickening on radial walls called casparian thickenings.

Next to the endodermis is the pericycle. It is represented by patches (cap like) of sclerenchyma present immediately above each vascular bundle.

The vascular bundles are many in number, arranged in the form of a ring surrounding a central pith. Each vascular bundle is conjoint (xylem and phloem combining to form a single bundle), collateral (xylem and pholem on the same radius with x/p arrangement), open (open to secondary growth i.e. cabium is present) with endarch (protoxylem pointing towards the pith) xylem. Phloem consists of seive tubes companion cells and phloem parenchyma . In between phloem and xylem is present cambium which is said to be intrafascicular (present within a vascular bundle). The cambial cells are thin walled, rectangular and actively dividing. Next to the cambium, internally is the xylem, with the protoxylem facing the pith. Xylem consists of vessels, trachieds, wood fibres and xylem parenchyma . Between the vessels are found the wood parenchyma.

Between the vascular bundles are present radially elongated parenchyma cells constituting the primary medullary rays. Occupying the central region of the stem is a wide parenchymatous pith.

STEM OF MAIZE

As an example of monocot stems, the stem of maize may be studied. A cross section of young stem shows an epidermis, a hypodermis and a ground tissue in which the vascular bundles are embedded (Fig 21) There is no clear cut distinction into stele , cortex and pith. Endodermis and pericycle are absent.

Epidermis is made up of a single layer of compactly arranged cells. It is covered with a cuticle. There are no epidermal hairs.

Hypodermis is internal to the epidermis. It consists of few layers of sclerenchyma cells.

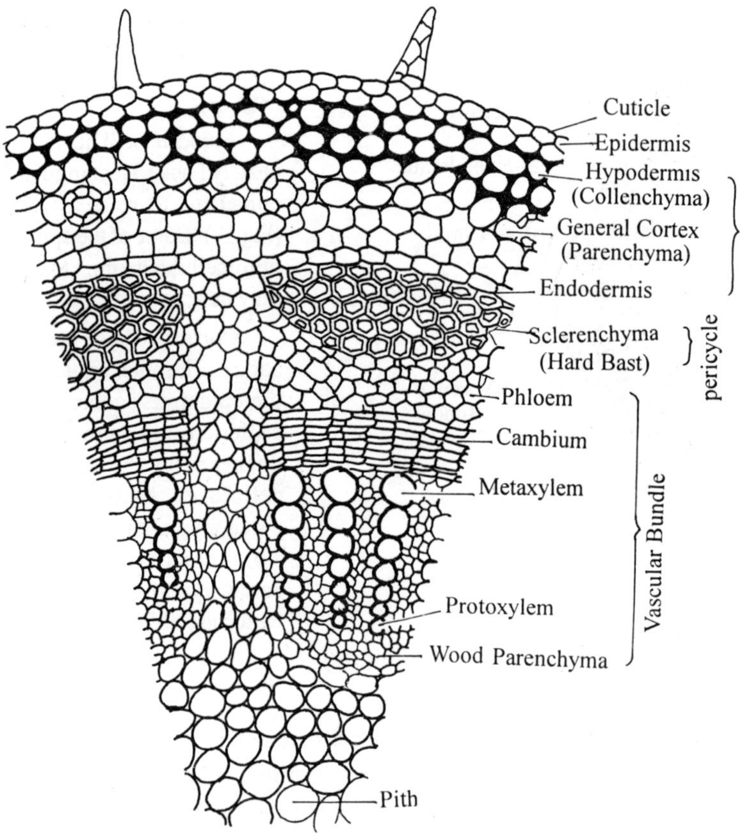

Fig. 13.4 T.S. of sunflower stem a part enlarged to show cellular details

Ground tissue is represented by loosely arranged, thin walled parenchyma cells. Embedded in this tissue are the vascular bundles. Vascular bundles are scattered in the ground tissue. They are crowded towards the margın but smaller in size, while towards the centre they are big in size and sparsely distributed.

Structure of a vascular bundle: Each bundle is oval in shape, conjoint, collateral endarch and closed (cambium is absent hence no secondary growth). Surrounding the bundle there is a sclernchymatous bundle sheath (Fig 22). The phloem is above the xylem. It consists of sieve tubes and companion cells. There is no phloem parenchyma . The outer cells of phloem (protophloem) are crushed while the inner cells (metaphloem) are healthy. Xylem is reduced; it consists of four (sometimes three) vessels (two meta and two proto) arranged in the form of 'Y'. Metaxylem vessels are pitted, while protoxylem vessels are annular and spiral. Sometimes in between the metaxylem vessels, small, pitted tracheas are present. Below the protoxylem a lysigneous cavity is present. It consists water and is known as the water containing cavity. Surrounding this are present cells of wood parenchyma .

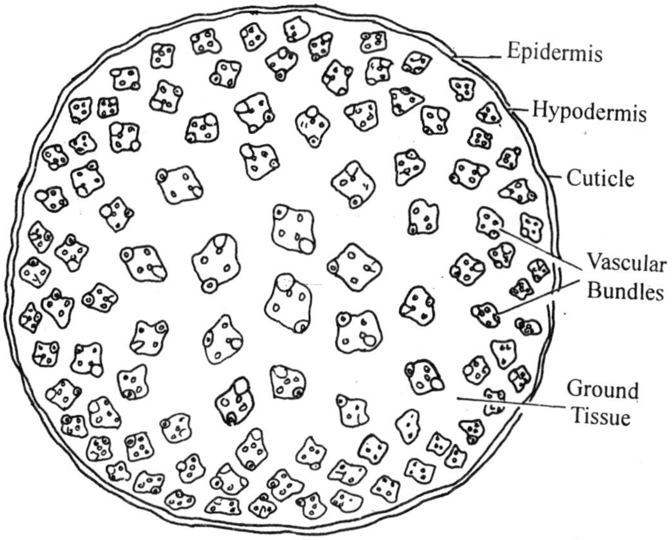

Fig. 13.5 T.S. of maize stem (ground plan)

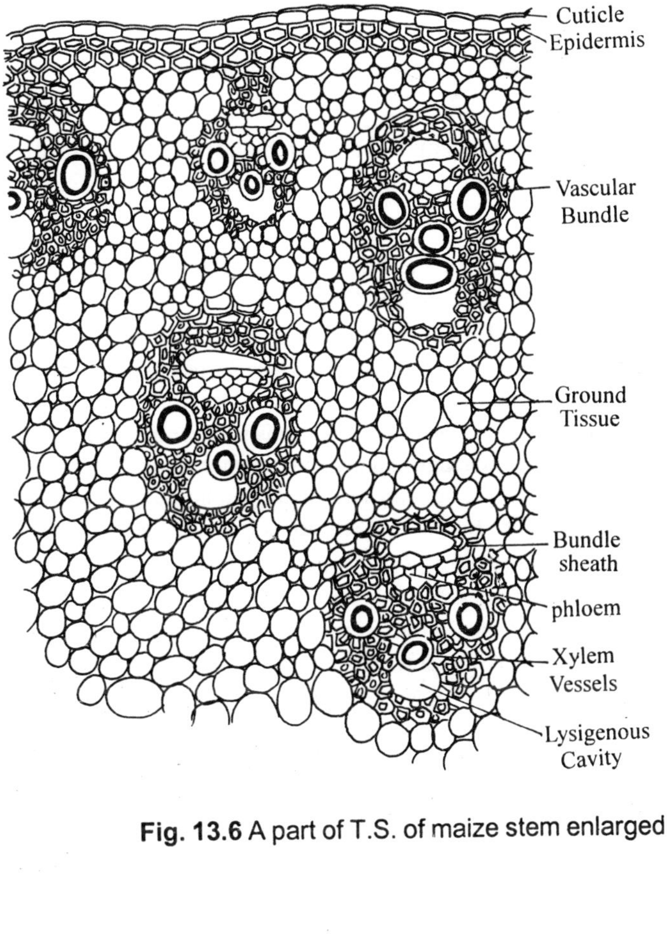

Fig. 13.6 A part of T.S. of maize stem enlarged

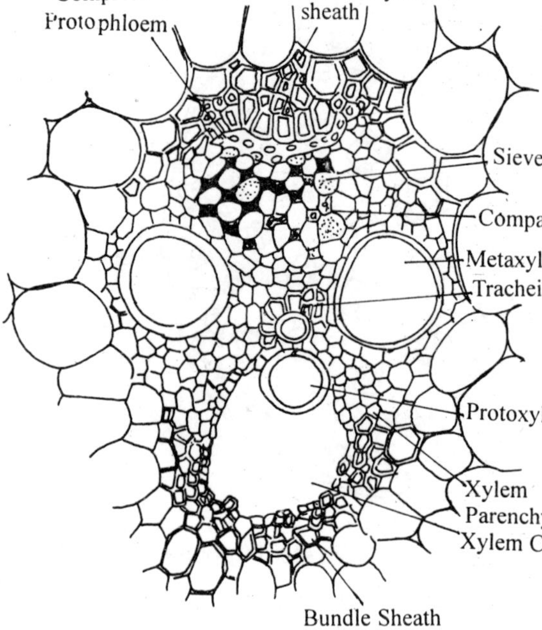

Fig. 13.7 A vascular bundle magnified

STUDY OF PLANT ANATOMY (continued): Isobilateral (Maize) and dorsiventral (Helianthus) leaves.

LEAF OF HELIANTHUS:

The leaf of sunflower shows the features of a typical dicotyledonous leaf. In a transverse section it shows three well defined regions viz. epidermis, mesoph· ll and veins.

Epidermis: There are two epidermal layers covering either surfaces of the leaf. The epidermis on the upper surface is called the upper epidermis and the one on the lower surface is called the lower epidermis. In some cases, covering the epidermal layers will be a layer of cuticle. In the leaf of sunflower the upper epidermis is continous while the continuity of the lower epidermis is interrupted by the presence of stomata. There are small pores helping in gaseous exchange and the loss of water during transpiration. Each stoma is covered by two guard cells which regulate the opening and closing of the stomatal aperture. The guard cells are crescent shaped with thick inner walls. Internally each stoma leads into an air space.

Mesophyll: In between the two epidermis layers is the mesophyll. It is differentiated and has well developed palisade tissue below the upper epidermis. Above the lower epidermis and below the palisade tissue is the spongy parenchyma which has loosely arranged cells with large number of air chambers. Palisade tissue is chlorophyllous and it is the chief photosyntietic tissue of the plant. There are a few chloroplasts in the spongy parenchyma cells also but their main function is storage. There are rich in starch grains.

Vein: The veins are arranged in the form of a network and in a cross section they show the vascular bundles. Each vein representing a vascular bundle is surrounded by a parenchymatous envelope called the bundle sheath. Within the bundle sheath are the vascular elements viz, the xylem and the phloem. Unlike in stems the xylem is above and the phloem is below.

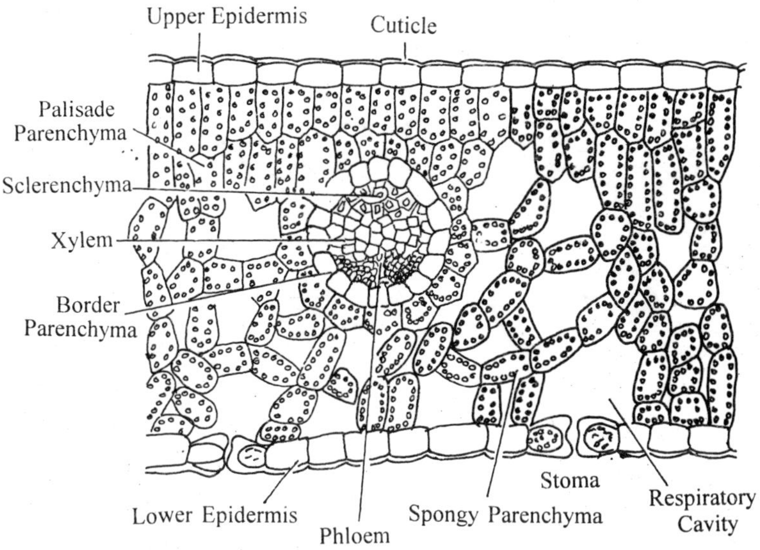

Fig. 13.8 T.S. of dicot leaf

LEAF OF MAIZE

As an example of a typical isobilateral leaf, the leaf of maize may be studied. A cross section shows the following (Fig. 24).

EPIDERMIS : There will be two epidermal layers upper and lower. The epidermal layers are uniseriate and composed of more or less ovel cells arranged compactly. There is the deposition of a cuticle both on the upper epidermis as well on the lower epidermis . Besides the normal cells, the upper epidermis consists of special cells called bulliform cells or motor cells. These cells are larger than the other epidermal cells and are thin walled. These help in the rolling of the leaves during dry weather . Leaves are amphistomatic (stomata are present on both the epidermal layers), Each stoma internally leads into a large substomatal air chamber.

Mesophyll: It is undifferentiated. But the mesophyll cells below the upper epidermis may show a compact arrangement.

Vascular bundles: The veins are arranged parallely and are not interconnected. There are two pads of sclerenchyma supporting the vein. The vein itself is surrounded by the bundle sheath made up of parenchyma cells. Within the bundle sheath are present collateral vascular elements showing x/p arrangement.

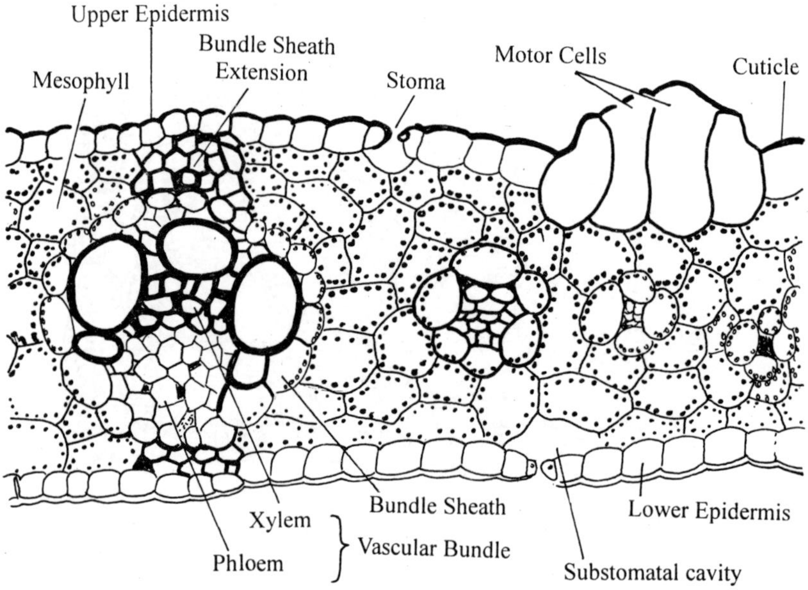

Fig. 13.9 T.S. of maize leaf

Table I Differences between Monocot stem and Dicot stem

Structure	Monocot stem	Dicot stem
1. Cortex	Made up of only parenchyma called the ground tissue. It spreads right up to the centre.	Made up of collenchyma, parenchyma and in some cases chlorenchyma also.
2. Endodermis	Absent	Present
3. Pericycle	Absent	Present
4. Vascular	Scattered and closed bundles	Arranged in the form of a ring open.
5. Cambium	Absent	Present
6. Pith	Absent	Present

Table II Differences between Moncot root and Dicot root

Structure	Monocot root	Dicot root
1. Xylem	Many bundles, exarch	2 to 6 in number exarch.
2. Pith	Wide pith is present	Either very narrow or absent.
3. Pericycle	Gives rise to lateral roots and cambium	Gives rise to lateral roots only.
4. Cambium	Absent	Present

Table III Differences between Monocot leaf and Dicot leaf

Structure	Monocot leaf	Dicot leaf
1. Mesophyll	Undifferenitated. Has loosely arranged parenchyma cell.	Differentiated into palisade and spongy parenchyma.
2. Sclerenchyma	Two pads of sclerenchyma present attaching the vascular bundle to the upper and lower epidermis.	A group of parenchyma cells present above the vascular bundle.
3. Upper epidermis	Motor cells (bulliform cells) present	Motor cells are absent.

CHAPTER 14

PLANT PHYSIOLOGY I

PLANT WATER RELATIONS

1. Physical demonstration of Osmosis (Thistle funnel experiment)

Materials. Egg membrane, animal bladder or parchment paper, 5% sugar solution, thistle funnel, beaker, water, stand etc.

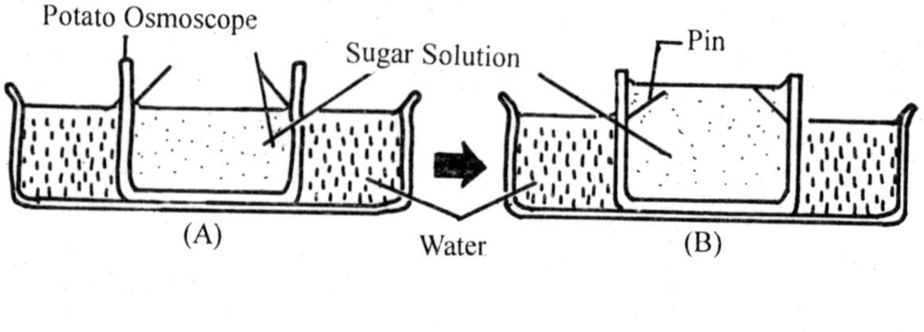

Fig. 14.1 Experiment to demonstrate physiological process of osmosis. Level of sugar solution before (A) and after (B) osmosis

Procedure : Take a beaker half filled with water. Take a thistle funnel and tie its mouth with egg membrane, parchment paper or animal bladder. Fix the thistle funnel to a stand inversley so that the funnel is immersed in water. Fill the funnel with 5% sugar solution. Note the level by making a mark on the stem of the thistle funnel. Allow the experiment to stand for about 30 minutes. Then note the final level of sugar solution. Taste the water in beaker.

Observation. The level of sugar solution in the thistle funnel has gone up. Water in the beaker does not taste sweet.

Conclusions. The egg membrane is selectively permeable only to water and not to sugar. Hence, as the solvent concentration is higher in beaker than in the thistle funnel, water migrates into the thistle funnel and the level of sugar solution rises up. This phenomenon is known as osmosis which may be defined as migration of solvent molecules from the region of their higher concentration to the region of their lower concentration across a selectively permeable membrane.

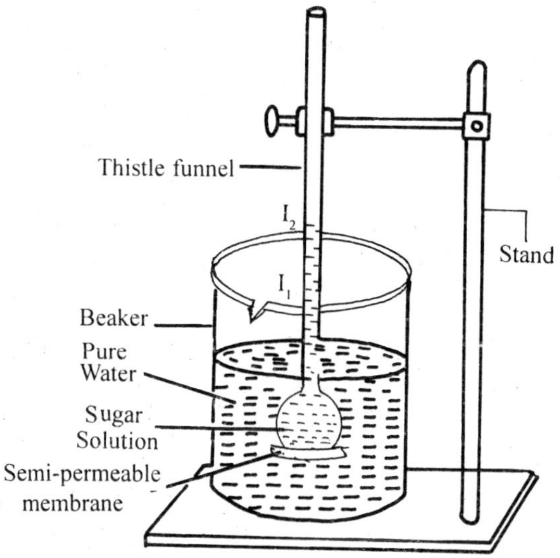

Fig. 14.2 Experiment to demonstrate physiological process of osmosis

2. Physiological demonstration of osmosis (Potato osmoscope)

Materials. Potato tubers, 5% sugar solution, water, petridish, pins.

Procedure. Select a large size raw potato and carefully peel of the skin. Cut into two pieces horizontally. Scoop out potato tissues from the centre and fashion a small cup as shown in the figure. Pour a little of sugar solution into the potato cup. Stick a pin inisde the cup to note down the level of sugar solution. Keep the potato cup (it is now called a potato osmoscope) in the petri dish containing water. Leave the apparatus aside for about 30 minutes. At the end of this period note the final level of sugar solution in the potato osmoscope and stick a pin to makr the level.

Observation. It is observed that the level of the sugar solution has gone up inside the potato cup due to entry of water from petridish.

Conclusion. Potato tissues act as a selectively permeable barrier and allow only water but not solute particles to pass through. As a result water which is in higher concentration in the petri dish. This migraion is due to osmosis. (For definition of osmosis see the previous experiment).

This experiment is known as physiological demonstration of osmosis as it is seen in living tissue, while the thistle funnel experiment is physical demonstration because it is seen in membrane removed out of the living tissues.

3. Demonstration of Ascent of Sap (Balsam plant experiment)

Purpose: To demonstrate Assent of Sap.

Requirements: Beaker, Water, Eosin and Balsam plant. Take some water in a beaker. Add a little eosin to it so that the water becomes red coloured.

Procedure- Cut a portion of the balsam plant and keep it in the beaker. Leave for sometime.

Observation: After some time red lines or streaks appear on the stem indicating that coloured water has moved up. These streaks represent the tissues through which water has moved up.

Remove the stem from the beaker and cut cross sections of the stem and observe under the microscope. On observation it will be seen that only xylem vessels are coloured.

Inference: Since colour water has moved up through only xylem, the path of ascent of sap must be xylem.

Note: Adding eoisn to the water is only to colour it so that the movement may be clearly observed.

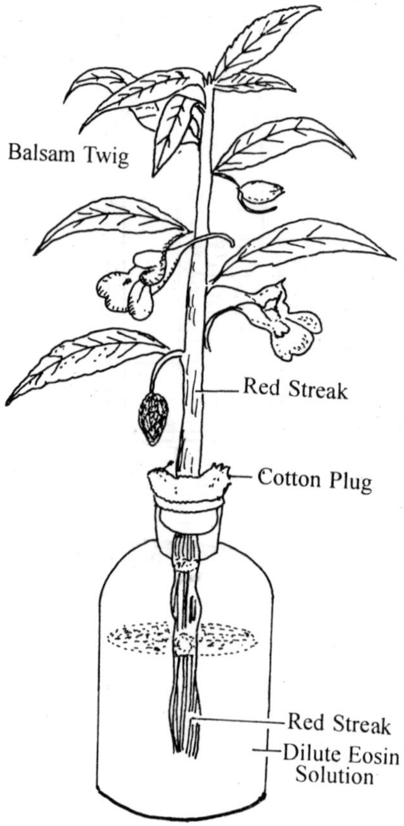

Fig. 14.3 Balsam plant experiment

4. Experiment to Demonstrate Transpiration pull:

The upward movement of water in the plant is due to the force of transpiration at the aerial parts. This evaporation of water creates a sort of a tension at the aerial parts and since the water molecules form a continuous stream within the plant, the tension is transferred down to the region of absorption and this lifts up the water. This force is called transpiration pull and this is held mainly responsible for the ascent of sap.

The phenomenon of transpiration pull can be demonstrated with the help of a simple experiment.

Materials required:

1. A freshly cut twig with leaves.
2. Graduated glass tube, rubber, thread, sealing wax, knife etc.,
3. Beaker, Mercury, Water, a suitable stand with clamps.

Procedure:

1. Take a suitable length of a leafy twig (cut under water) and insert the twig into a graduated tube with the help of a split cork.
2. Fill the tube with water. Close the lower end with thumb and make it air tight by sealing wax after tying it with a thread.
3. The glass tube which is now air tight is completely filled with water is carefully held with finger and immersed into a beaker containing mercury. The entire experimental set up is held upright with the help of a stand and clamp.
4. Keep the experimental setup in bright sunlight for about 1 or 2 hours.
5. The mercury column will rise in the tube which can be recorded for determination of suction force.

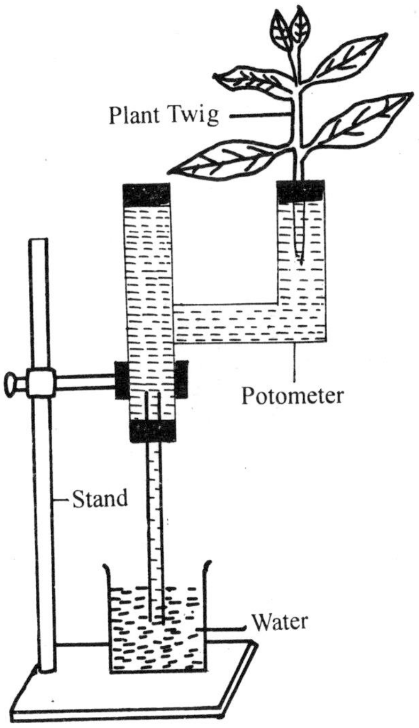

Fig. 14.4 Experiment to demonstrate transpiration pull

Observations: The rise of mercury through the graduated tube indicates the suction force or the transpiration pull. As the transpiration goes on, the force created by it will lift the water column in the glass tube and subsequently the mercury also is forced up in the tube.

Conclusion: This suction force is due to transpiration.

5. Demonstration of transpiration by bell jar method

Materials. A healthy potted plant, glass plate, vaseline, oil, cloth or polythene bag, thread etc.

Procedure. Select a healthy potted plant and keep it on a glass plate. Cover the top of the pot with a polythene bag or oil cloth leaving a hole for the plant. Tie the cloth securely to the pot. Cover the pot with a bell jar and apply vaseline to the edges of the bell on the glass plate to make it air tight. Now keep the entire set up in sunlight. Observe the apparatus about 30 minutes.

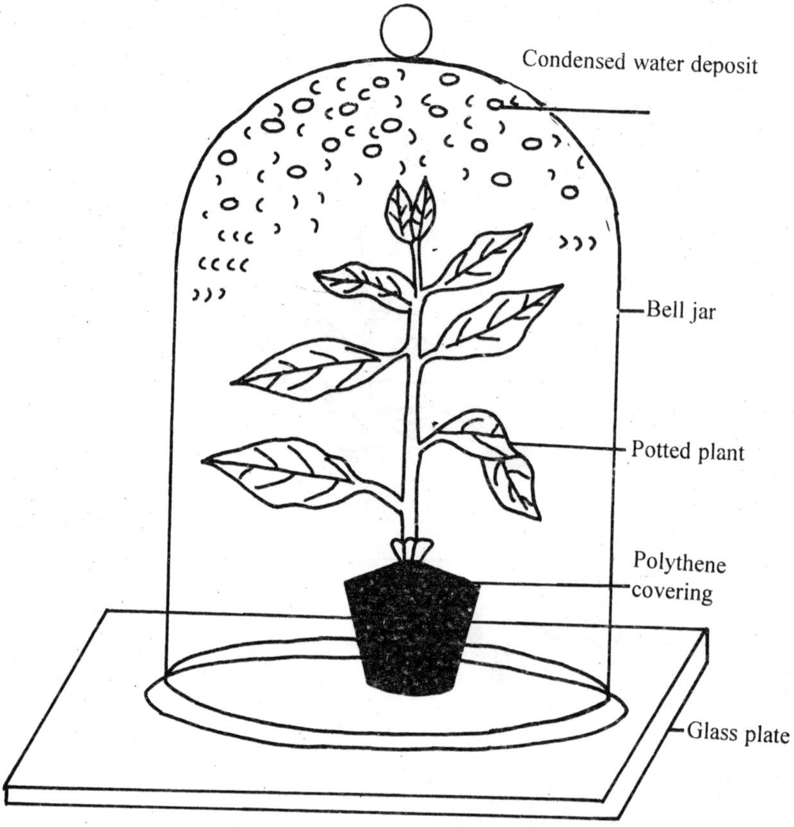

Fig. 14.5 Experiment to demonstrate transpiration

Observation. Small drops of water appear on the inner side of the bell jar.

Conclusion. This experiment demonstrates transpiration. As the water vapours come out of the plant they get condensed on the walls of the bell jar and form water drops.

6. Measurement of rate of transpiration by Ganong's potometer

Materials. Ganong's potometer, beaker, plant, eosin, water.

Procedure. The apparatus consists of a horizontal graduated tube which is bent in opposite directions at either ends. One bent end is wide and the other end is narrow A reservoir is fixed to the horizontal tube near the wider end. The reservoir has a stop cock to regulate the water flow. The aparatus is filled with water by opening the reservoir. A twig or a small plant is fixed to the wider arm through a split cork. The other bent end of the horizontal tube is dipped into a beaker containing water coloured with eosin. An air bubble is introeuded into the graduated tube at the narrow end (arm). As transpiration takes place, the air bubble will move from one end of the graduated tube to the other. After it has reached the reservoir, it (reservoir) is opened to push the air bubble back and then its movement can be watched agian.

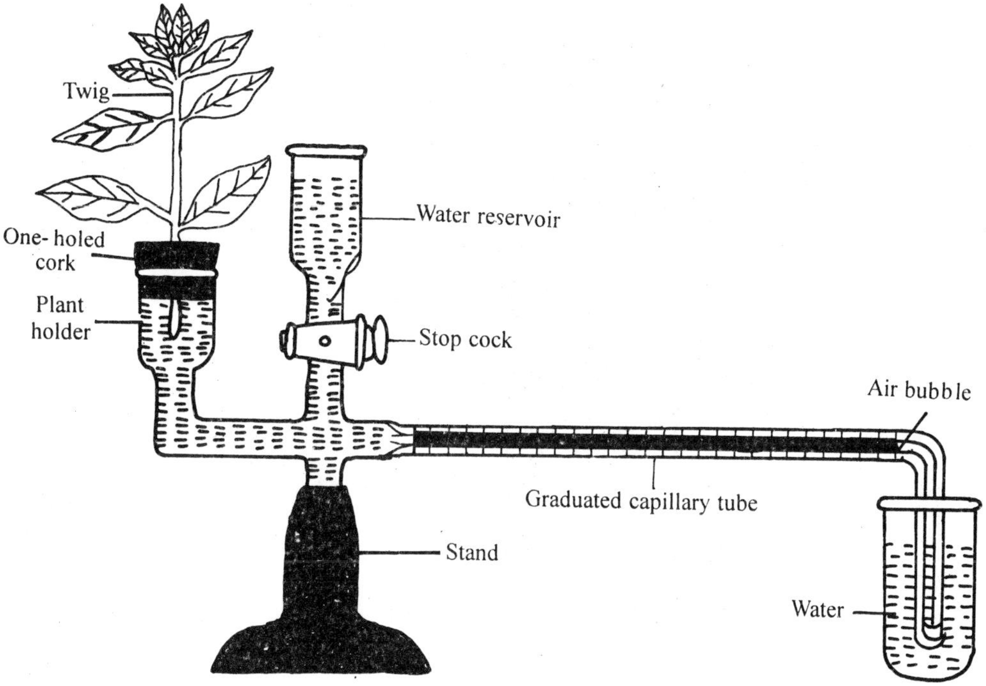

Fig. 14.6 Ganong's potometer

Observation. Record the time that is needed for the air bubble to move from one end to other. Make many observations to obtain the average of time taken to travel the distance.

Conclusion. The Ganong's potometer measures the rate of transpiration indirectly by measuring the rate of absorption. The rate of transpiration is approximately equal to the rate of absorption. By using the following formula the amount of water transpired may be calculated.

$V = \Pi r^2 1$

Where V = volume of water, Π = 22/7, r = radius of graduated tube, 1 = distance travelled by the air bubble.

7. Measurement of transpiration by Farmer's potometer

Materials. Farmer's potometer, beaker, water, vaseline, plant.

Procedure. The Farmer's potometer consists of a wide mouthed bottle. A cork with three holes is fixed on to the mouth of the bottle. One whole is fixed with a thrice bent tube with a scale fixed on to it horizontally, the second whole has a reservoir with a stopper filled onto it and through the third whole fix a twig cut under water.

Fill the apparatus with water. Open the reservoir so that the tube gets filled with water. Make the apparatus air tight by applying vaseline. Dip the free bent end of the horizontal tube into a beaker containing water. Introduce an air bubble into the tube. Keep the apparatus in sunlight for some time.

Observation and conclusion. Same as in previous experiment.

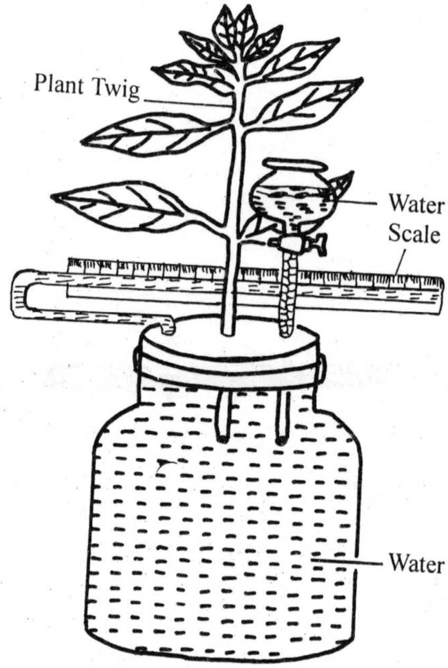

Fig. 14.7 Farmer's potometer

CHAPTER 15

PLANT PHYSIOLOGY II

BIOENERGETICS

1. Evolution of O_2 during photosynthesis

Materials. Beaker, water, glass funnel, test tube, *Elodea* or *Hydrilla* plant, pryrogallol.

Procedure. Take a glass beaker half filled with water and put some pieces of *Hydrilla* or *Elodea* plant into it. Invert a glass funnel over it. Fiaall a test tube with water and carefully invert it over the stem of the funnel so that the mouth of the test tube is under water. Keep the apparatus in sun light and observe.

Observation. Small air bubbles appear from the cut ends of branches of the plants and they enter the test tube. Gradually water in the test tube gets downwardly displaced and it (test tube) becomes filled with the gas. The gas can be tested by one of the following methods.

A. Cover the mouth of the test tube and carefully take out the test tube. Introduce a burning splinter into the test tube. It glows brightly.

B. Dissolve some amount of pyrogallol in water. The gas is dissolved in pyrogallol and the level of the water in the test tube rises again.

Conclusion. Both the tests performed above will show that the gas is oxygen. The aquatic plants kept in the beaker photosynthesize in the presence of sun light and release oxygen which comes out of the plant in the form of air bubble.

2. Chlorophyll is necessary for photosynthesis

Materials. Plant with variegated leaves such as *Codeum, Coleus, Manihot* etc, butter paper, pencil, 70% alcohol, potassium iodide solution, water bath.

Procedure. Select a potted plant with variegated leaves and keep it in darkness for 72 hours so that it gets destarched. Bring out the plant and place it in bright sunlight. Select a leaf with variegated colours and placing a butter paper on the leaf trace the outline of various colours with the help of a pencil. After few hours pull ou the leaf from the plant, boil it in 70% alcohol in a water bath so that the leaf gets decolourized. Pour KI solution on the leaf. Observe the change in colour.

Observation. Blue colour appears only in those areas which were green. This can be checked with the outline diagram of the leaf.

Conclusion. Chloroplyll pigment is necessary for photosynthesis. Only the green areas in the leaf photosynthesize and starch is present in those regions. That is why when KI solution is poured they turn blue (indication of starch). Other areas (non green) do not have strach and do not turn blue on contact with KI solution.

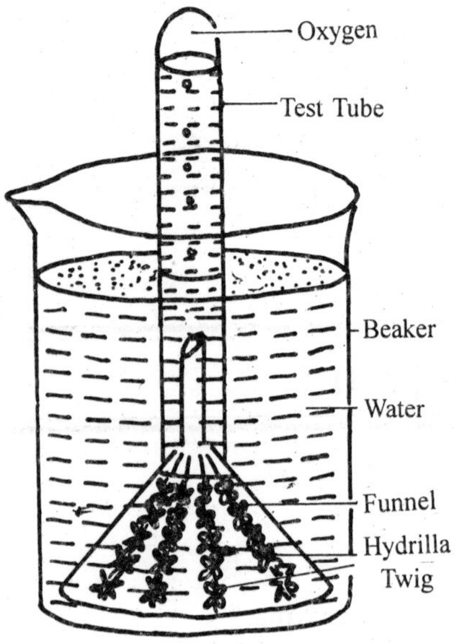

Fig 15.1 Apparatus used to demonstrate evolution of O_2 during photosynthesis

3. **Light is necessary for photosynthesis (Ganong's light screen experiment)**

Materials. A potted plant, Ganong's light screen, beaker, 70% alcohol, KI solution, test tube, water bath.

Procedure. Keep a potted plant in darkness for about 72 hours so that it is starch free. Fix a Ganong's light screen to one of the leaves and place the pot in bright sunlight. After few hours pluck the leaf from the plant, remove the light screen and decolourize it by boiling (in a test tube having 70% alcohol) it in a water bath. Take out the decolourized leave and pour KI solution.

Observations. The area of leaf covered by light screen remains bleached while the other areas turn blue on contact with KI solution.

Conclusion. The area covered by the light screen does not get light hence starch is not manufactured in that area. As a result it does not turn blue on contact with KI solution. Other areas get light so they manufacture starch and turn blue on contact with KI solution.

4. **CO_2 is necessary for photosynthesis (Mohl's half leaf experiment)**

Materials. Wide mouthed bottle, split cork, KOH, healthy plant.

Procedure. Keep a potted plant in darkness for about 72 hours to remove starch. Select a leaf from such a plant and without removing it insert half a portion into a bottle through a split cork. Before inserting the leaf pour some quantity of KOH liquid into the bottle. Make the apparatus air tight by smearing vaseline to the edges of the cork. Keep the apparatus in sunlight for about 2 hours. At the end of the period remove the leaf from the bottle detach it from the plant and boil it in 70% alcohol to decolourize it. Pour Iodine solution and observe the change in colour.

Observation. The part of the leaf inside the bottle will remain bleached whereas the portion outside turns blue.

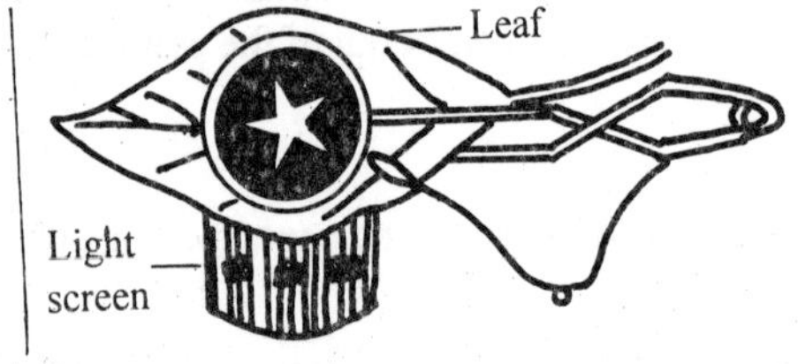

Fig 15.2 Ganong's light-screen

Conclusion. The portion of leaf inside bottle had all the necessary factors for photosynthesis except CO_2 because it was absorbed by KOH. The portion of leaf outside the bottle turns blue as it has starch manufactured by photosynthesis. It was provided with all the factors including CO_2. This demonstrates clearly that CO_2 is necessary for photosynthesis.

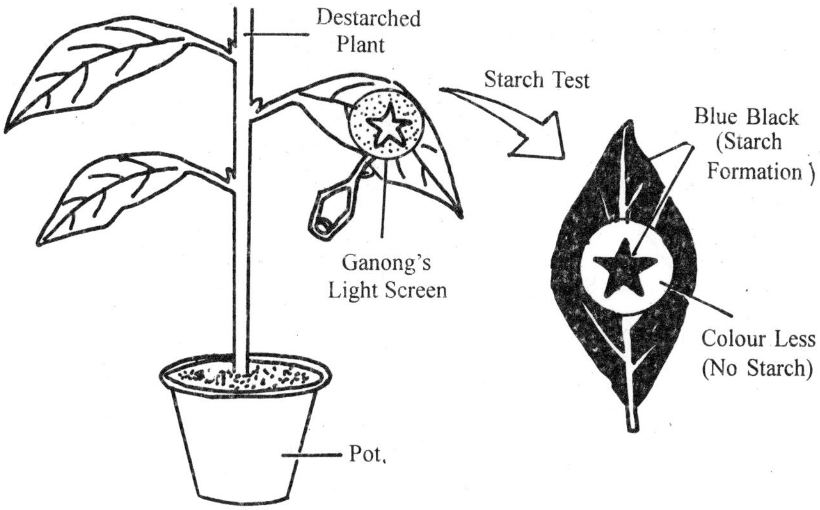

Fig. 15.3 (A) Experiment to show the necessity of light for photosynthesis. (B) Leaf tested for starch

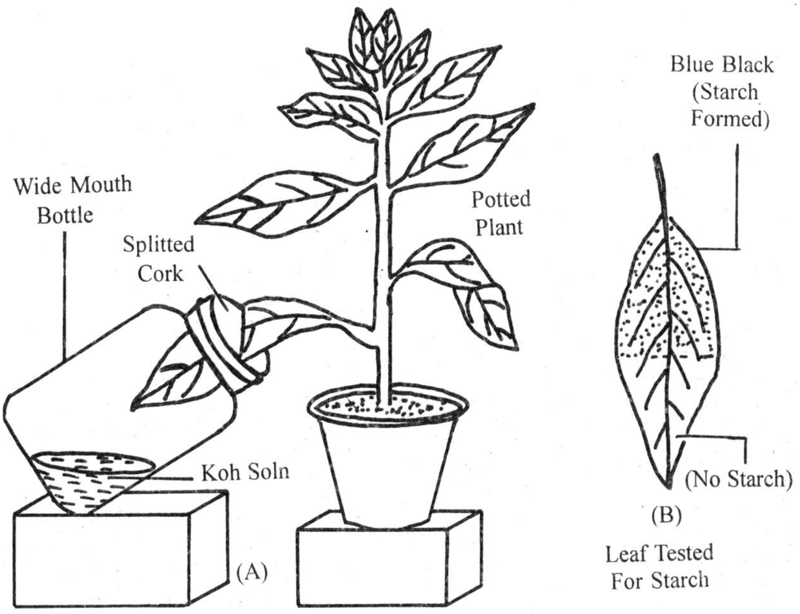

Fig. 15.4 (A) Experiment to show that CO2 essential for photosynthesis (B) Leaf tested for starch

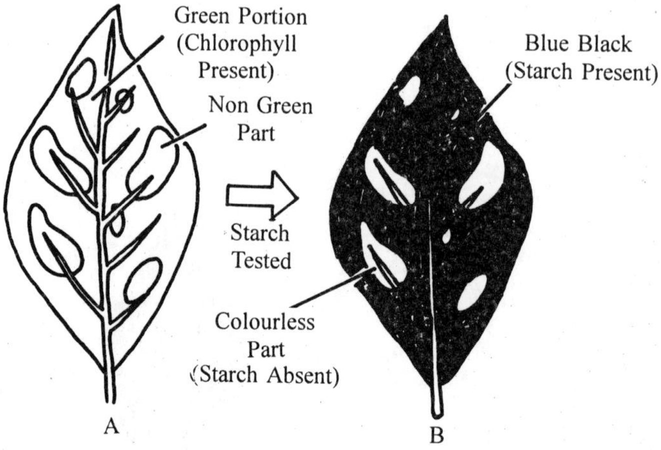

Fig. 15.5 Test for starch in a variegated leaf

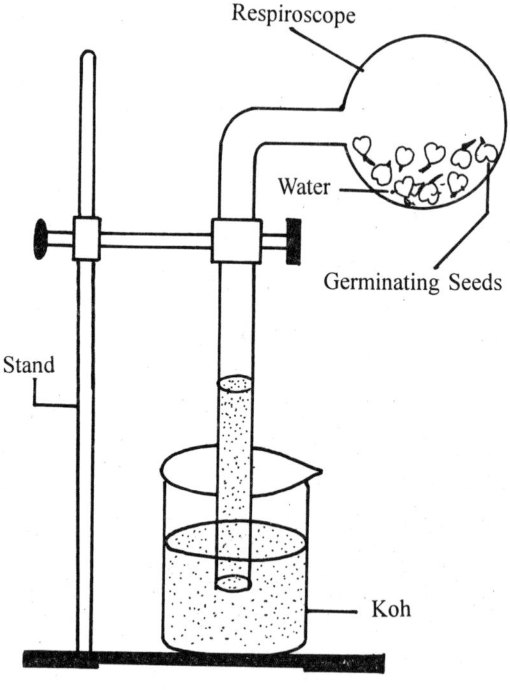

Fig 15.6 Demonstration of evolution of CO_2 using simple respiroscope

5. Release of CO_2 during aerobic respiration

Materials. Conical flask, a twice bent tube, test tube, thread, germinating seeds, KoH solution, vaseline, beaker, saline water.

Procedure. Take a conical flask and place some germinating seeds in it. Close the mouth with a single holded cork. Introduce a twice bent tube through the hole. Take a small quantity of KOH solution in a test tube and hang it inside the flask with the help of a thread. The thread will be held tight with the help of the cork. Let the free end of the bent tube be dipped in a beaker containing saline water. Apply vaseline to the cork to make it air tight. Let the apparatus stand for somtime.

Observation. Water level in the tube rises after sometime.

Conclusion. As the germinating seeds respire they release CO_2 which is absorbed by KOH. This creates a vacuum in the tube making saline water to rise.

6. Evolution of heat during respiration

Materails. Two thermos flasks, germinating seeds, mercuric chloride, rubber cork etc.

Procedure. Take two thermos flasks. Keep germinating seeds in one and mercuric chloride treated (killed) seeds in the other. With the help of single holed corks introduce a thermometer into each of these flasks and close the mouth (of the thremos flask). Note the initial temperature in both the flasks. After 45 minutes note the temperature again. Record your observations.

Observation. In the thermos flask with killed seeds there will be neither increase nor decrease in temperature as recorded by the thermometer. In the other flask having germinating seeds there will be an increase in temperature as recorded by the thermometer.

Conclusion. During respiration there is release of energy, while most of it used for the synthesis of ATP, some amount escapes as heat and this results in increased temperature in the flask with germinating seeds. In the other flask where the seeds are killed, there is no increase in temperature as there is no respiration.

7. Experiment to demonstrate the evolution of CO_2 using Ganong's respiroscope

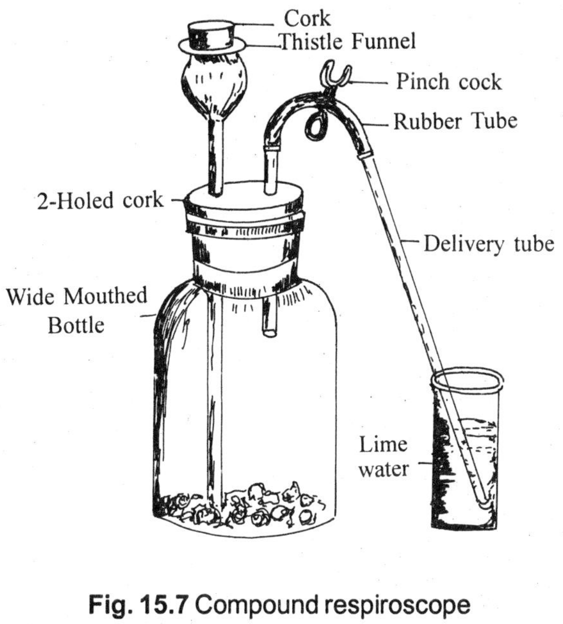

Fig. 15.7 Compound respiroscope

Materials. Ganong's respiroscope, germinating seeds, beaker, stand and KOH solution.

Experiment. The Ganong's respiroscope consists of a vertical tube which is built on one side to the bent end attached a bulb. The other end of the tube should be dipped in the beaker containing a solution of Potassium hydroxide (KOH). Before the beginning of the experiment some germinating seeds are kept in the bulb of the respiroscope.

Observation. After sometime it may be noticed that the level of the KOH has risen in the tube.

Conclusion. The germinating seeds release CO_2 during respiration. This CO_2 is absorbed by the KOH solution and hence rise in its level.

8. Experiment to demonstrate the volume of CO_2 evolved and oxygen utilised during aerobic respiration (double respiroscope experiment)

Materials. Double respiroscope, germinating seeds, KOH solution and water

Experiment. Fix the two respiroscopes on a vertical stand as shown in the figure. Germinating seeds are kept in both the respiroscopes. In the first respiroscope a small test tube containing KOH solution is hung inside with the help of a thread the free end of which is attached to the stopper on top. The second respiroscope will not have the tube with KOH solution. Close the stopper at the upper end of the respiroscopes. The lower ends of both the respiroscopes are dipped in beakers containing water.

Observation and conclusion. After some time it is seen that there is rise in the level of water in the first respiroscope as the carbondioxide produced is absorbed by the KOH solution creating a vacuum to fill which water rises up. In the second respiroscope there is no rise in the level of water as no vacuum is created.

Further in the second respiroscope the volume of CO_2 evolved is equal to the volume of O_2 absorbed so that the net volume of air inside remains same. In the first respiroscope the rise in the level of water incidates the amount of CO_2 released.

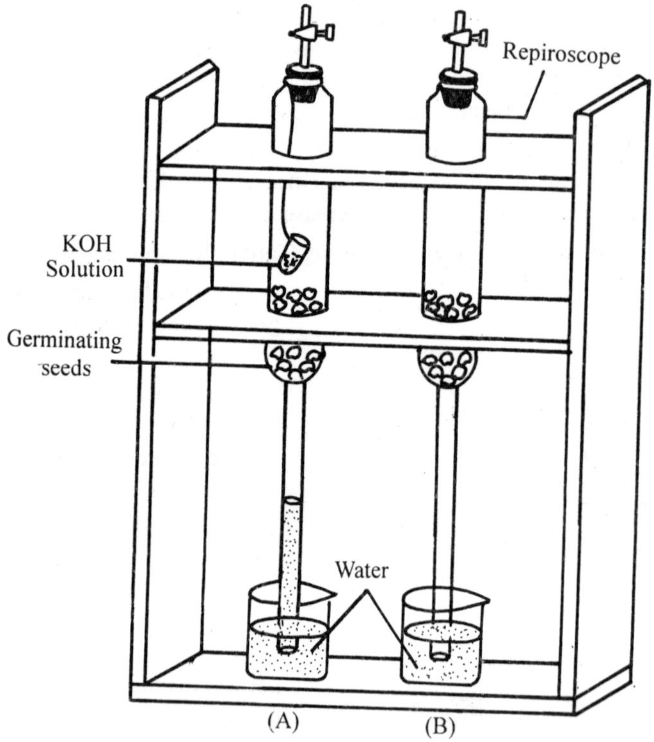

Fig. 15.8 Double respiroscope to find out R.Q.

9. Ganong's Respirometer

Purpose - To determine the value of respiratory quotient (RQ) by using different respiratory substrates.

Apparatus- Gangong's respirometer consists of the graduated tube connected to a glass bulb and a leveling tube. Both these tubes are connected with the help of a rubber tubing. The entire apparatus is fixed vertically to a stand (see fig 39). The glass bulb has a stopper. The neck of the bulb has a small rounded hole. A similar hold is found on the side wall of the glass stopper. When the stopper is fixed to the bulb both the holes will be in same line and this will allow the entry of air into the bulb. Both the tubes are half filled with either mercury or saline water (pure water is not used for the experiment as it can absorb carbon dioxide).

Procedure - Open the mouth of the bulb by taking out the stopper. Introduce some wet cotton into the bulb and place it at the bottom. Next, keep some germinating seeds on the wet cotton. The wet cotton provides moisture necessary for the continued germination of the seeds. Close the stopper and slightly rotate it so that the two holes come to lie in the same line. As a result of this the air inside the bulb is in direct communication with outside. Adjust the levelling tube so that the level of saline water in both the tubes will be at the same level. Now rotate the stopper so that the two holes will be out of line and the air in the bulb is not in communication with the exterior. At this stage note the level of the saline water in the graduated tube. Keep the apparatus for; sometime note the reading.

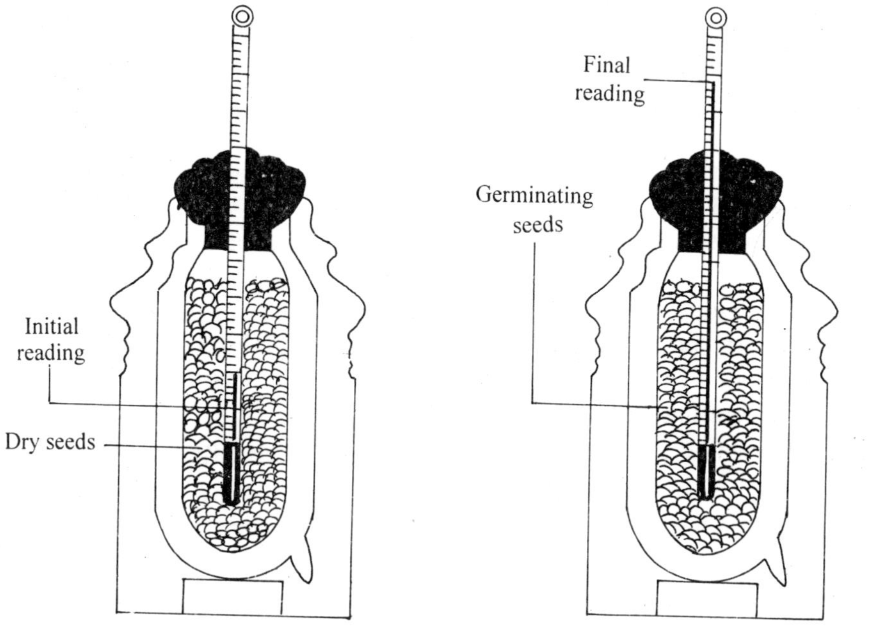

Fig. 15.9 Experiment to show that heat is generated in respiration.

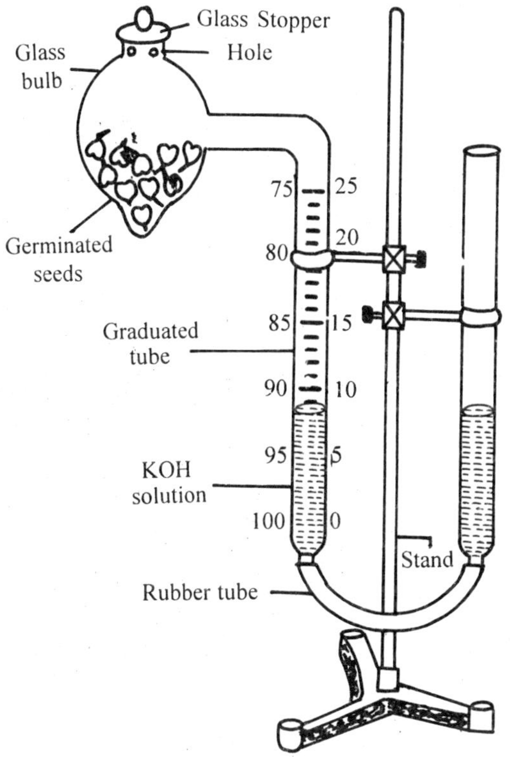

Fig 15.10 Ganong's respirometer

Observation and conclusion : The level of saline water in the graduated tube will not be altered if the germinating seeds (as in cereals like wheat, oats etc) have carbohydrates as the stored food. This is because the amount of carbon dioxide released will be equal to the amount of oxygen used. Hence the level remains unchanged.

In case the germinating seeds have fat as the stored food (castor, mustard or linseed etc) the level in the graduated tube when compared with the levelling tube will rise. This is because when fat is oxidised more oxygen is absorbed as compared to the O_2 released. Therefore volume of gases in the bulb will decrease and the level is increased.

10. Demonstration of anaerobic respiration

Materials. Petri dish, test tube, germianting seeds, mercury, KOH crystals, forceps etc.

Procedure. Fill a test tube with mercury and invert it over a petri dish also containing mercury. Now lift the test tube lightly and introduce few germinating seeds (gram, pea etc with testa removed) with the help of a foreceps. The seeds will rise to the top of the test tube. Keep the test tube back in its position and allow the apparatus to stand for about 45 minutes.

Observation. A gas starts building up at the top of the test tube and it pushes the mercury down. Now introduce a piece of caustic potash. This absorbs the gas and the level of mercury goes up.

Conclusion. Germinating seeds have a high rate of respiration. As they are kept in mercury they do not get oxygen and hence respire anaerobically. Since introduction of KOH absorbs the released gas it (gas) must be CO_2. This experiment demonstrates that CO_2 is released during anaerobic respiration.

11. Demonstration of fermentation (Kuhne's fermentation flask)

Materials. Kuhne's flask, yeast extract or toddy, 5% glucose solution, KOH pellets.

Procedure. Kuhne's flask consists of a slightly slanted glass tube with a side bulb attached to it. It has a basal stand. The slanted tube has no opening at the top while the side tube with a bulb has an opening which can be closed with a stopper.

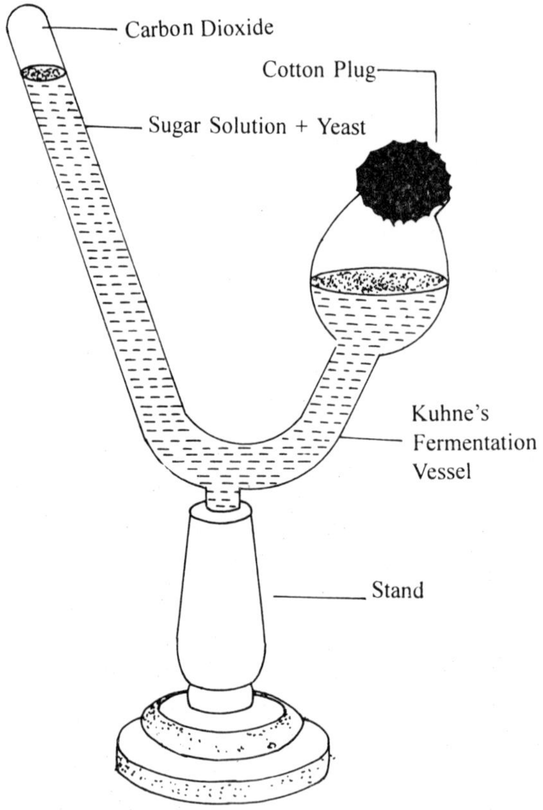

Fig 15.11 Kuhne's fermentation vessel

Fill the Kuhne's flask with 5% sugar solution in such a way that the slanted tube is full but the bulb is only filled up to half its capacity. Add yeast extract or few drops of toddy into the glucose solution. Close the opening of the bulb with a stopper or with cotton. Let the apparatus stand for sometime and observe.

Observation and conclusion : Small bubbles of gas start rising in the slanted tube and push the level of the solution below with the result a gas filled space is created at the top. If this is allowed to continue the glucose solution fills the side bulb and it may overflow if there is no stopper. Introduce a piece of KOH into the tube through the bulb so that it reaches the gas filled top, the gas is absorbed and the glucose solution once again moves to the top of the tube. The gas must be CO_2 as it is absorbed by KOH. This shows that CO_2 is released during fermentation. This is different from anaerobic respiration because here the respiratory substrate is outside the cell (glucose solution). Alcohol is formed during fermentation.

GROWTH

12. Measurement of growth by Arc auxanometer

Materials. Arc auxanometer, weight, thread, a potted plant.

The apparatus. The arc auxanometer consists of a tripod stand and a pulley connected to a pointer which moves on an arc scale. A thread is made to pass over the pulley tightly. One end of the pulley is tied to a weight.

Procedure. Take a healthy potted plant and tie its tip to thread which is held over the pulley in such a way that the thread does not slacken. Keep it aside and observe.

Observation and conclusion. As the stem tip grows it will push the thread up and the tight thread will turn the pulley. The pulley in turn will move the pointer over the graduated arc scale. The extent of mc vement from the initial to the final in a given period of time can be measured. The distance moved by the pointer does not give the actual linear growth, because the arc is divided into 90° (one quarter of a circle ie., 360/0.25) and not linear measurements. Further the measurement on the graduated arc is magnified because of the length of the pointer. The actual growth is calculated as follows:

Initial position of pointer = A

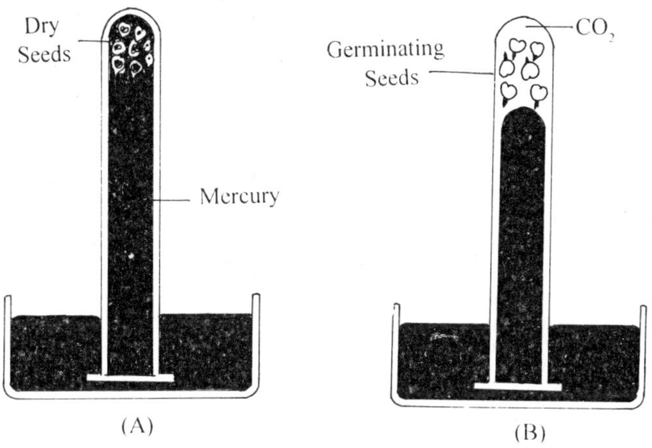

Fig 15.12 Demonstration of anaerobic respiration

Final position of pointer = B

Actual distance travelled by the pointer during a given period of time = B-A.

Actual growth = B-A X radius of pulley in length

Length of pointer (from the centre of the pulley)

For example if the radius of pulley is 10cms and the length of the pointer is 50cms the magnification is 10 times. If the pointer has moved 5 cms in 20 hours the actual growth is as follows:

$$\frac{5 \times 10}{50} = 1\text{cm in 20 hours.}$$

Therefore growth per hour 1/20 = 0.5mm.

QUADRAT METHOD OF STUDYING VEGETATION, Calculation of relative density of a species.

A quadrat means a square. In plant ecology when the vegetation has to be studied in the field with reference to the species composition, species density, relative frequency etc, it will be difficult to survey the entire area as it may be very vast. For this purpose quadrats of suitable sizes are laid in different places and the species present there are analysed and the data obtained is projected for the entire area.

In the following experiment, a quadrat method is followed to calculate the relative density of a particular species.

Requirements : Four nails or pegs, string, graph sheet, herbarium sheet etc.

Laying a Quadrat: Using the four nails or pegs and the string create a square of one meter. This can be done by fixing the nails or pegs at suitable distances and then tying a string to enclose all of them. First identify all the species present within the quadrat and then carefully count the number of individuals in each species and then calculate the relative density of any particular species using the following formula.

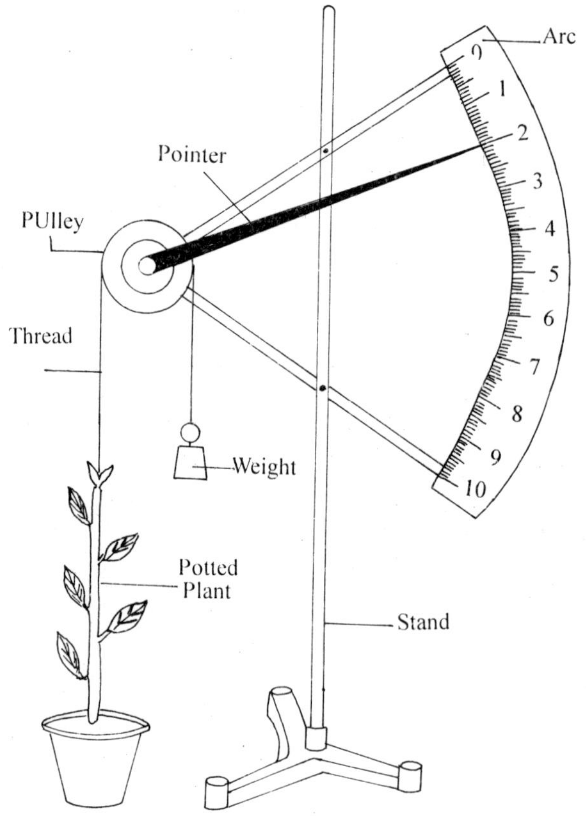

Fig 15.13 Arc auxanometer

$$\text{Relative density of species A} = \frac{\text{Number of individuals of species A}}{\text{Total number of individuals of all species}} \times 100$$

Analysis of data : The number of individuals of a given species when compared with the total number of individuals of all the species will give an idea of its density. This value multiplied by hundred gives the percentage density. A sample example of the species present in a quadrat and their densities is given below.

Species symbol	Name of the Species	Total individuals	Relative Density
A	*Synedrella*	25	13.8
B	*Euphorbia*	36	19.8
C	*Emilia*	70	38.6
D	*Indigofera*	50	27.6

CHAPTER 16

PROTOZOA

AMOEBA

For the purpose of the laboratory study amoeba culture are obtained from pond water, decaying vegetative etc.

1. Amoeba is unicellular. It changes its shape continuously.
2. The size varies from 0.2 - 0.3 mm in diameter.
3. There is no cell wall. The plasma membrane is the outermost layer.
4. Internal to the plasma membrane is the cytoplasm. It can be divided into an outer layer called Hectoplasm and an inner region called Endoplasm.
5. Lying in the endoplasm is the single large nucleus.
6. There is also a contractile vacuole situated in the endoplasm. It helps is excretion and osmoregulation.
7. Present in the endoplasm are a number of food vacuoles containing food particles.
8. Amoeba moves with the help of finger like projections called Pseudopodia.
9. Reproduction in amoeba is generally by either binary fusion or multiple fusion.

Systematic Position

Phylum	:	Protozoa
Class	:	Rhizopoda
Order	:	Labosa
Type	:	*Amoeba proteous*

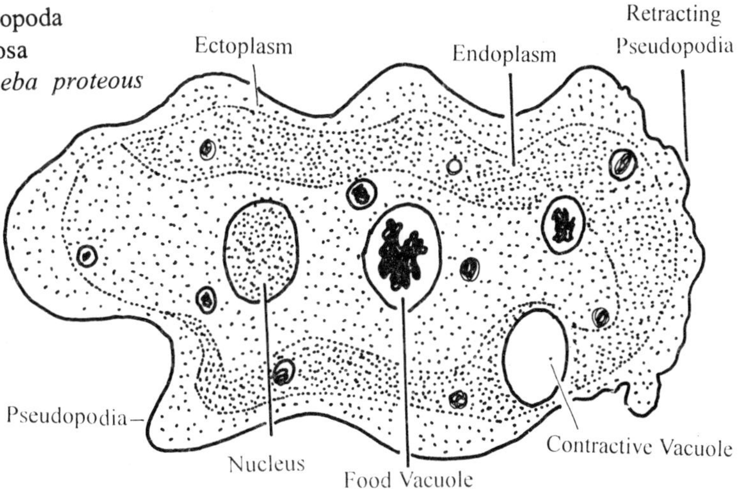

Fig. 16.1 Amoeba

PARAMECIUM

Paramecium is popularly called 'slipper animalcule' because of the shape of the body. It occurs abundantly in fresh water ponds which are rich in decaying animal and vegetable remains. A drop of pond water freshly collected will show under the microscope these animlacules moving to and fro in the filed (of the microscope).

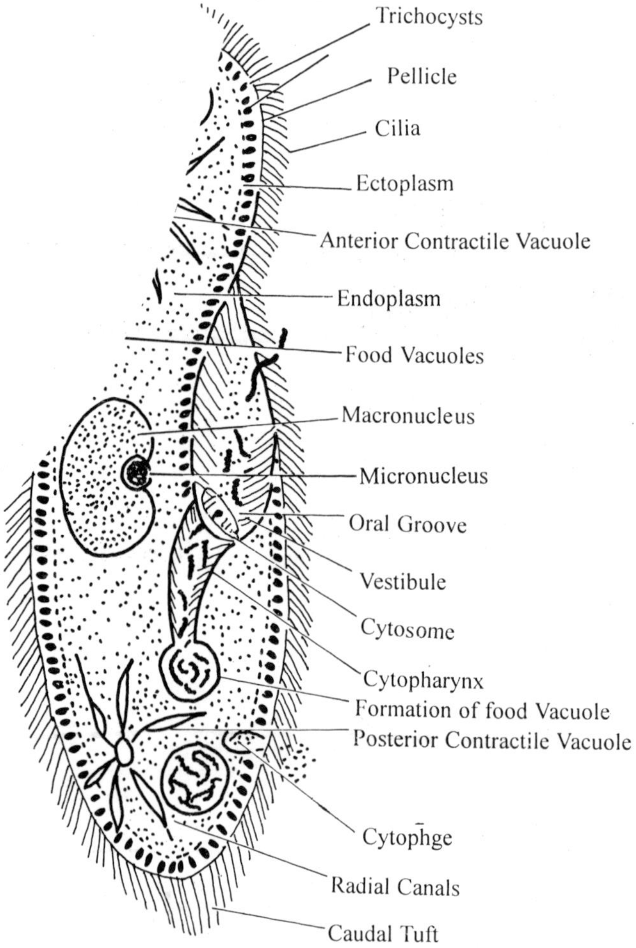

Fig. 16.2 Paramecium

The body of Paramecium is made up of a single cell about 0.3mm in length and is shaped like the sole of a slipper hence the name slipper animalcule. One end which is somewhat narrow is called the anterior end and the broad end is called the posterior end. The anterior end is in front during locomotion. One of the flat surfaces which always faces the substratum and on which the animal glides is called the oral or ventral surface while the opposite end is known as the aboral or dorsal surface. Present in the oral surface near the anterior end is an oral groove. The oral groove leads to the mouth (of the animal) called cytosome. The mouth opens into a gullet or cytopharynx which passes obliquely downwards pointing to the posterior end. Usually food vacuole is always seen towards the end of the gullet.

Externally the body is covered by a cell membrane to which are attached a number of small hair like structures called cilia. These are the locomotory organs. The animal moves with the help of cilia and in addition, the cilia also help in gathering food particles due to their lashing movement. The cilia are arranged in longitudinal rows which run lengthwise in the hinder part of the body but follow the spiral twist on the oral surface. Two to three rows of cilia lining the gullet are somewhat larger than those present on the general surface. They are fused to form an undulating membrane which is attached to the dorsal wall of the mouth.

Internal to the membrane (pellicle) is the cytoplasm divided into two regions-ecotplasm and endoplasm. Ectoplasm is the outer clear non granular region while endoplasm is the inner region made up of a granular mass.

The membrane, also called the pellicle is thin but rigid enough to maintain the shape of the cell. It also affords support to all the internal motons of cytoplasm. The pellicle is made up of a number of hexagonal depressions with a hole in the centre. Cilia emerge out through these holes. These are thin protoplasmic process which have their origin in basal granules. The ciliary movement is regulated by a neuroneme or neuromotor system which is a sensory fibre running all along the pellicle parallel to it. Present in the dorsal ectoplasm are two contractile vacuoles occupying either end of the cell. Each contractile vacuole consists of a central vesicle and a number of radiating canals which open into it. The whole structure appears star shaped.

The ectoplasm just below the layer of cilia consists of a large number of sacs alternating with the basal granules. These sacs called trichocysts are spindle shaped, 1/100 mm in size and are filled up with a colourless fluid of high refractive index. When paramecia are subjected to some external stimulus or irritation the trichocysts are ejected out through the openings present in the pellicle ridges. During discharge, trichocysts become elongated (about eight times their normal length). The function of these trichocysts is not certain.

Endoplasm is granular and less viscous than the ecotplasm. The most prominent component of endoplasm is the nuclear apparatus consisting of a macronucleus (large) and a micronucleus (small). The micronucleus lies in a cleft on the side of the macronucleus. While the macronucleus regulates all the cellular activities, micronucleus seems to be concerned with conjugation (reproduction).

Also present in the endoplasm are a number of food vacuoles. In some strains of Paramecium, a number of wedge shaped particles called kappa particles are found in the endoplasm.

Systematic Position

Phylum	:	Protozoa
Class	:	Ciliata
Order	:	Holotricha
Type	:	*Paramecium caudatum.*

VORTICELLA

1. Vorticella belongs to the class Ciliata.
2. It is found in fresh water and attached to certain objects.

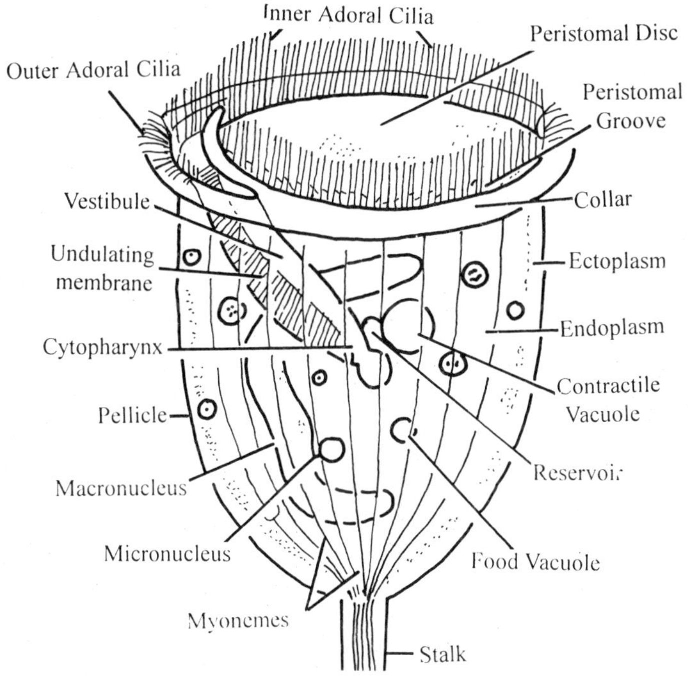

Fig. 16.3 Vorticella

3. The body is made up of an inverted bell like structure, below the bell there is a stalk.
4. Near the wide portion of the bell there is a rim called the collar.
5. Inside the collar is found a peristomial groove . This groove surrounds a centrally situated peristomial disc.
6. The peristomial groove is capable of contracting, when it contracts the peristomial disc is drawn inwards.
7. The peristome leads to another groove infundibilum which leads to the cytopharynx.
8. The tip of the cytopharynx ends in a food vacuole.
9. The outer covering of the body is called the pellicle.

10. There are parallel lines called myonemes present below the pellicle region. These help in the contracting of the body organism.
11. The cytoplasm can be divided into hectoplasm and endoplasm.
12. Macro and micro nucleus are found in the endoplasm.
13. The stalk is the lowermost portion and is also capable of contraction.
14. Reproduction vorticella is by means of binary fusion as well as by sexual reproduction.

Systematic Position

Phylum	:	Protozoa
Class	:	Ciliata
Order	:	Peritricha
Type	:	*Vorticella companula*

EUGLENA

Euglena is a fresh water free moving organism. The entire body is made up of a single cell. It is elongated and spindle shaped. One end is somewhat blunt and the other end is somewhat tapering. The blent end is the anterior end. It has a flask shaped cavity which opens to outside through cytostome followed by cytopharynx. The cavity itself is called the reservoir. Arising from the reservoir is the long whip like structure - the flagellum. The flagellum helps in locomotion. The flagellum arises from two granules called basal granules. The body is covered by the means of the pellicle. There is a single large nucleus in the cytoplasm. Also distributed in the cytoplasm are a number of chloroplasts which help the organism in photosynthesis. Situated near the reservoir is the large contractile vacuole. Surrounding this are found a number of small contracting vacuoles. Situated also near the reservoir is the stigma or eye spot which is photoreceptive.

Euglena reproduces by binary fusion.

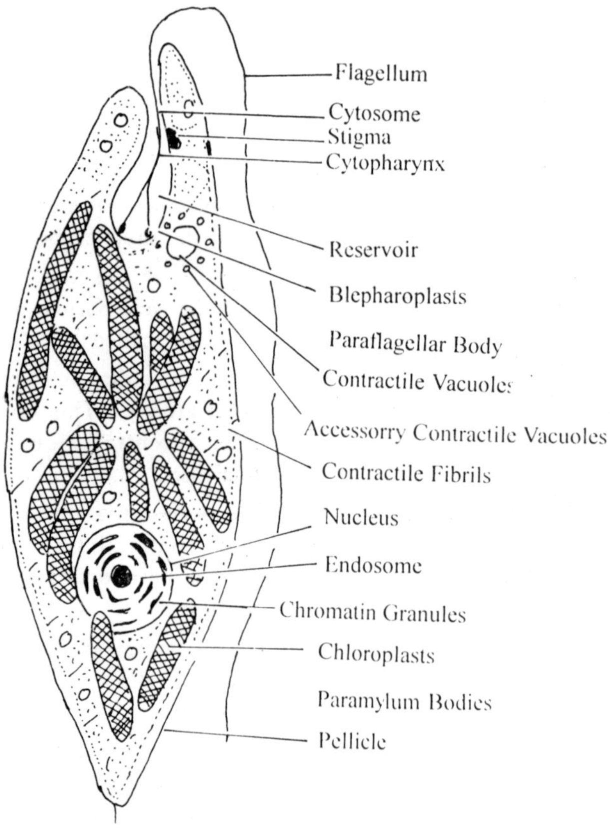

Fig. 16.4 Euglena

Systematic Position

Phylum	:	Protozoa
Class	:	Mastigopthora
Sub - class	:	Phyto-mastigophora
Order	:	Euglenida
Type	:	*Euglena virdis*

CHAPTER 17

CELL AND CELL DIVISION

A smear of buccal epithelium should be made on a thin glass slide, stained properly and observed into the microscope. The animal cell does not have cell wall. It has only a cell membrane as the outermost covering. Within the cell membrane is the cytoplasm. The cytoplasm has nucleus and many other inclusions like centrosome, golgi apparatus, mitochondria, ribosomes etc.

Definition of Mitosis

The division of the body or somatic cell is known as Mitosis. It is a complex process involving series of changes leading to the formation of two equal daughter cells from a single parent cell. It was first observed and named by Fleming in 1882. It is otherwise called somatic cell division, as it occurs in all body or somatic cells

Process of mitosis. In mitosis the nuclear division precedes the cytoplasmic division. The former is called the Karyokinesis and the later cytokinesis. When cytokinesis does not follow karyokinesis, it results in multinucleated cells. Mitosis may be divided into 5 significant stages or phases.

1. Interphase
2. Prophase
3. Metaphase
4. Anaphase and
5. Telophase

Interphase

It is actually the interval between two mitotic divisions when no visible changes occur in the cell. But during this significant period important metabolic activities of nucleus and cytoplasm take place. This results in the growth of the cell. Interphase is longest in the cell cycle. The characteristic features of this phase are :

1. Nucleus appears conspicuously large.
2. Nucleolus is highly enlarged.
3. DNA and protein synthesis takes place.
4. The chromosomes are feebly visible as extremely thin strands or threads. They are in duplicated or replicated state due to the replication of DNA molecules.

Prophase

1. Chromosomes coil (spiralisation) and condense and become visible as thread like strands. The duplicated members of a chromosome are known as chromatids. The number of chromosomes is constant for a species.

By the end of prophase they move to the equatorial centre of the cell.

2. The nucleolus gradually reduces in size and disappears by the end of prophase. Its material becomes deposited on the chromosomes.

3. The nuclear membrane breaks down and dispenses into the cytoplasm.

4. The centrosome with the centriole divides forming amphiaster. Astral rays appear around each centriole forming asters. As the two asters separate and move apart spindle fibres begin to form between them.

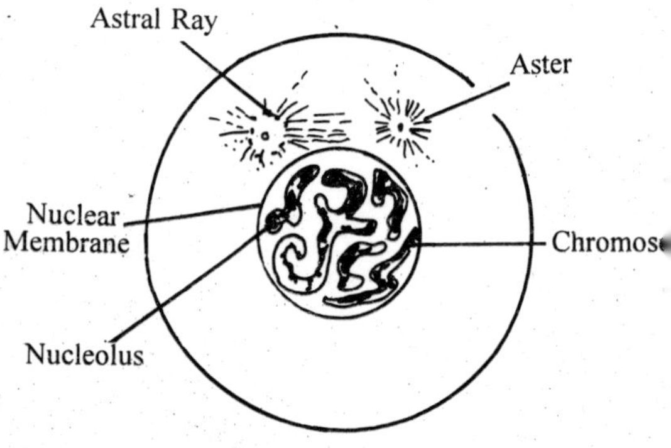

Fig. 17.1 Prophaase

Eventually the amphiaster constituting the two polar asters and spindle fibres establishes the 'poles' and 'places' of division.

In most plant cells there are no centrioles. But 'poles' are organised at the opposite ends of the cell. Spindle fibres are formed between the two poles.

Metaphase

1. In early metaphase the highly shortened and condensed chromosomes show characteristic radial arrangement on the equatorial plane of the spindle. This pattern of orientation of the chromosome is known as 'metaphase plate'. It is formed at right angles to the spindle axis.

2. The duplicated chromatids or daughter chromosomes are still united by the single centromere. The centromeres are attached to the spindle while the arms of the chromatids are free facing the opposite poles.

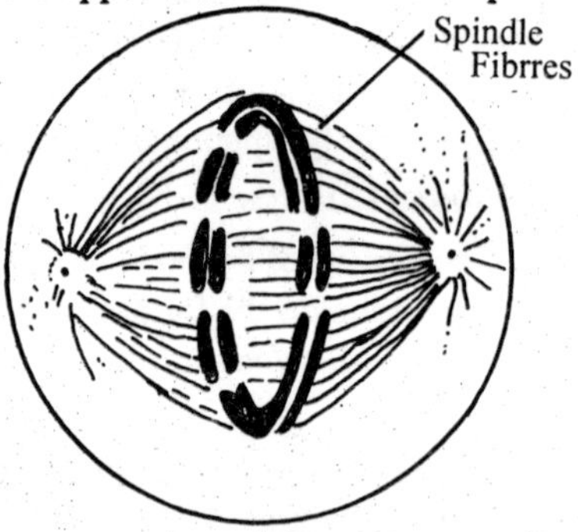

Fig. 17.2 Metaphase

y the end of metaphase the single centromere replicates and each chromatid comes to possess its own centromere.

Anaphase

1. With the replication of centromere, the sister chromatids separate and migrate to the respective poles. The sister chromatids are now called chromosomes.

2. Three types of spindle fibres are distinct in this stage. The spindle fibres which are attached to the centromeres are known as chromosomal spindle fibres. The spindle fibres which connect the two polar asters are known as continuous spindle fibres. The spindle fibres which form in between the departing chromosomes are called as the interzonal spindle fibres.

3. Equal number of chromosomes are pulled towards the opposite poles by the contraction of chromosomal fibres, by the stretching of interzonal fibres and other cytoplasmic events.

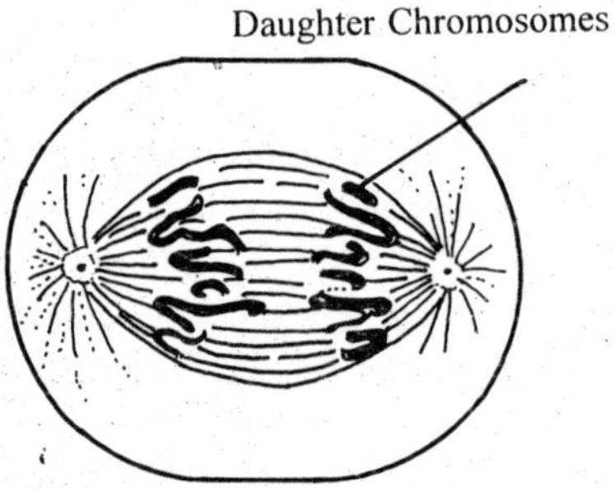

Fig. 17.3 Anaphase

4. During the polar migration of chromosomes, they appear as 'rod' shaped (acrocentric), 'L' shaped (submetacentric) and 'V' shaped (metacentric) structures.

5. In this manner the genetic material of the parent cell is equally distributed to the daughter cells.

6. End of the anaphase is marked by appearance of clusters of chromosomes at each pole of the cell.

Telophase

When the chromosomes reach their respective poles, reorganisation of daughter nucleus and cytokinesis sets in. This differs in plant and animal cell. In plant cells cytokinesis does not occur simultaneously with telophase as in the animal cell.

1. In plant cell as the nuclear membrane reappears, a cell plate is formed at the equator of the spindle. In due course it separates the two daughter cells.

2. In animal cells a 'furrow' appears at the equator and expansion of the plasma membrane takes place. A 'contracting ring' forms at the furrow, constricting the cytoplasm in the centre. Finally the cytoplasmic mass is divided into two surrounding each nuclei. In both animal and plant cells, the spindle fibres disappear.

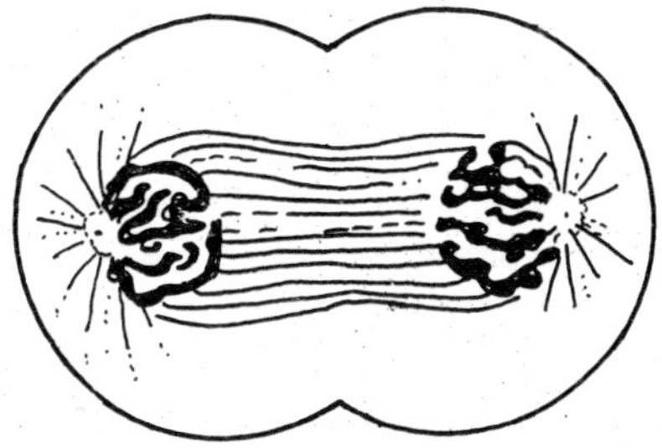

Fig. 17.4 Telophase

3. In animal cells the asters lose their fibres and reconverted to centrioles. The centrioles take up their natural position outside the nucleus.

4. The chromosomes uncoil into their strands and become indistinct.

Thus, two daughter cells are formed from the parent cell. The new cells are genetically identical with each other and the parent cell. Soon they enter into interphase during which metabolic, synthetic and growth activities of the cell take place.

Significance of mitosis

1. Mitosis is an extremely regular process. In maintains the quantitative and qualitative distribution of hereditary materials to the daughter cells.

2. It maintains constancy of the diploid number of chromosomes of species.

3. It maintains the constancy and balance in the DNA and RNA content of the cell.

4. It maintains the nuclear and cytoplasmic volume in the cell.

5. It results in the growth and development of the organisms.

6. Repair and replacement of dead cells is ensured by the mitosis.

7. It plays an important role in asexual reproduction and in the formation of gametes in sexually reproducing organisms.

In higher plants and animals, sexual reproduction takes place by the union of specialised sex cells and 'sperms' and 'ova'. In the fusion of two sex cells what prevents the doubling of chromosome in each new generation? The answer lies in a special type of nuclear division called Meiosis.

Process of meiosis. It involves two cell divisions. Meiosis I or heterotypic division and Meiosis II or **Homotypic division.** Each division is further divided into Prophase, Metaphase, Anaphase and Telophase

Meiosis I includes Prophase I, Metaphase I, Anaphase I and Telophase I.

The meiocyte which enters the prophase I goes through an interphase during which growth and doubling of genetic material and cytoplasmic components takes place.

Prophase I is the longest phase of meiosis, when the most significant genetic and cytoplasmic changes take place. The nucleus enlarges remarkably. This phase is subdivided into Leptonema, Zygonema, Pachynema, Diplonema and Diakniesis, on the basis of nuclear behaviour.

Leptonema is characterised by the extremely thin and long uncoiled thread like chromosomes. They show the presence of seed like thickenings all along the length of chromosomal strands called chromomeres. The chromosomes are diploid in number. They may be polarised towards the direction of centrioles. Nucleolus enlarges in size. The centrioles form the amphiaster and start moving apart to the opposite poles.

Zygonema. Marks the attraction and paring of paternal and maternal chromosomes. As already described elsewhere, two chromosomes which are alike in size, shape, structure, composition and function are called homologous chromosomes. They are derived from the two parents. The pairing of chromosomes is called synapsis. The synapsised homologues are known as bivalents. They undergo condensation. The nucleolus enlarges more. The amphiaster moves apart forming spindle fibres in between.

Pachynema. In this stage the two chromosomes of a bivalent undergo longitudinal replication or duplication. The replicated members are styled as sister chromatids. As a result, each bivalent has now 4 chromatids. This condition is known as tetrad. Gradual condensation and shortening of bivalents takes place.

At this time, peculiar breaks occur at the same points of the 2 non-sister chromatids of a bivalent. It is at this point identical 'blocks of genes' are exchanged between the two and the broken ends fuse with each other. This phenomenon of exchange of genetic material between nonsister chromatids is known as genetic crossing over. The point of exchange appears as an 'X' shaped structure known as chiasma (chiasmata). The number of chiasmata varies depending on the length of the chromosomes. Chiasma is observed in most plant and animal cells.

The nucleolus enlarges to a maximum size. The amphiaster spindle formation progresses further.

Diplonema. The force of attraction which held the tetrads of the bivalents close together as a unit ceases and is replaced by the force of repulsion. This pushes apart the non-sister chromatids away from each other. This leads to the displacement of chiasmata. However at the point of chiasmata the homologues are still no contact. The process of repulsion of homologues and displacement of chiasmata is known as terminalisation.

Condensation and shortening of the bivalents continue and the distinctly enlarged nucleolus begins to break down. Nuclear membrane begins to disintegrate. Amphiaster is almost complete.

Diakinesis. The bivalents show maximum condensation thickening and shortening. The terminalisation is complete. They move towards the nuclear membrane. The bivalent chromosome appear X, +, O or loop shaped depending upon the number of chiasmata. The nucleolus and nuclear membrane disintrigate and disperse into the cytoplasm. Amphiaster is well established.

Metaphase

The bivalents in tetrad condition move along and attach themselves on the equator of the spindle. Their arrangement is peculiar and different from that of mitotic metaphase. Here centromeres attached to the spindle, face towards the opposite poles, while the arms, towards the equator. The force of repulsion between the homologues of the bivalents reaches maximum, separating the two. Thus in metotic metaphase the separation of homologous chromosomes is not caused by the replication of centromere as in mitotic metaphase.

Anaphase

It is marked by the separation of the bivalent tetrad into material diads (2 chromatids) and paternal diads. This separation of paternal and maternal diads is known as disjunction. Thus the resulting nuclei will have either the paternal homologue or maternal homologue but never both. This rescues the chromosome number from diploidy to haploidy. Hence the division is known as 'reduction division'.

Telophase

The polar groups of diad chromosomes unwind and become thin long and thread like. The nucleolus reappears. The nuclear membrane is formed again. Thus two daughter nuclei are organised having only haploid sets of diads. The aster loses its radiating rays to become centriole. The spindle gets dispersed. The karyokinesis is followed by the cytokinesis resulting in two equal haploid daughter cells. The cytokinesis does not lead to formation of equal daughter cells in the case of formation of a ova in the ovary.

Interphase

The interphase between the first and second meiosis is mostly very brief if not absent. In the former case the most important feature is the absence of DNA and chromosome replication. However, RNA and protein synthesis and other metabolic activities take place.

Meiosis II or Homotypic division

It is a slightly modified meiotic division. During this, the two haploid cells in turn divide into 4 haploid cells. The two members of diads are separated into monads and are shared equally between the resulting cells.

Prophase II

The two chromatids of the diads are still connected by the single centromere. With the loosening of the arms the diads appear as 'X' shaped structures. The nucleolus and nuclear membranes begin to disintegrate. The amphiaster makes its appearance.

Metaphase

The diads line up on the equator of the spindle. The centromeres replicate separating each diad into two monads.

Anaphase II

The separated monads are now called as chromosomes. Equal number of chromosomes move along the spindle towards the opposite poles.

Telophase

Arrival of the haploid chromosomes at the opposite poles marks the telophase. The two daughter nuclei are reconstricted with the appearance of nucleoli, nuclear membrane, reformation of centriole and disappearance of spindle.

Cytokinesis

It is immediately followed by which the cytoplasm divides equally around each haploid nuclei. It may be unequal during the formation of ova.

In this manner 4 haploid cells are formed from the original diploid cell in two successive divisions.

CHAPTER 18

INVERTEBRATA

SYCON

1. Scypha or Sycon possesses a vase-shaped and radially symmetrical body which measures from 1 to 3 cm in length.

2. The colour is not specific but varies from grey to light brown.

3. Near the attached end are found a few small tubular projections, or the buds, which are formed not so extensively as in Leucosolenia or other asconoid sponges.

4. The free end of the vase-shaped individual bears an opening, the osculum (exhalent or excurrent pore), which is fringed with long, straight, needle-like monoaxon spicules.

5. This oscular fringe checks small animals from entering into the body.

6. The body surface is thrown into regularly arranged polygonal elevations from which project separ-like spicules, called oxeotes (monaxon spicules) that impart a bristly appearance to the body.

7. The polygonal elevations are separated by deep grooves, bearing minute pores, ie., the ostia, which lead into central body cavity, the spongocoel or paragastric cavity, through a system of canals.

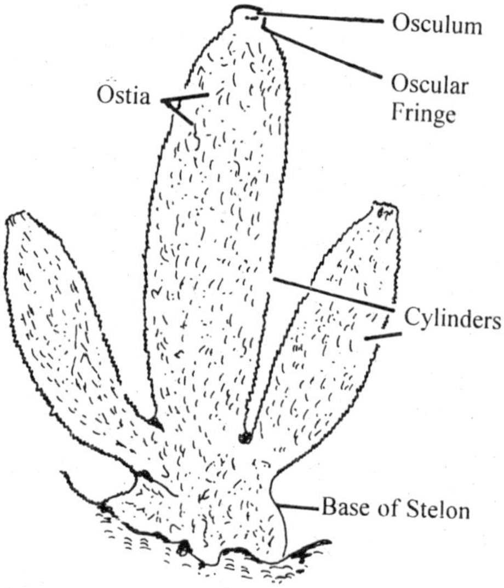

Fig. 18.1 Sycon

Systematic Position

Phylum	:	Proifera
Class	:	Calcarea
Order	:	Heterocoela
Type	:	*Sycon geiatinosum*

HYDRA

Hydra belongs to the phylum Colenterata. It is 2-20 mm in length and easily visible to naked eye. The body of hydra has the form of the tube. It's proximal end is closed by a flat disc called the basal disc or the foot which fixes the body to the substratum. The other end of the body has a small conical projection called hypostome. It has an opening - the mouth. The mouth is surrounded by a ring of 4 -12 finger like projections called tentacles. The central region of the body between the basal disc and the hypostome has a number of projections called as buds. These buds help in asexual reproduction.

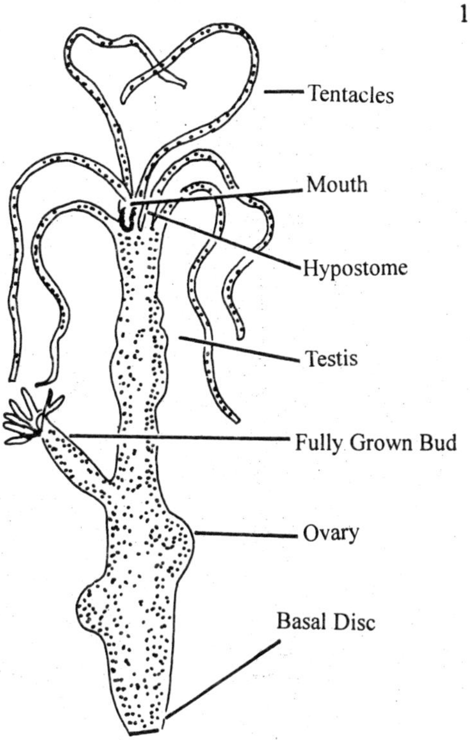

Fig. 18.2 Hydra

Systematic Position

Phylum	:	Coclenterate
Sub Phylum	:	Cnidaira
Class	:	Hydrozoa
Order	:	Hydroidea
Type	:	*Hydra virdis*

JELLY FISH

Aurelia has a flattened saucer shaped body. It measures upto 4 inches in size. The body looks like an umbrella and it has a convex exumbrellar side and the concave subumbrellar side. In the centre of the exumbrellar is the short and inconspicuous manubraum. The manubraum has a mouth at its free end which is squarish. From each end of the square hang down a finger like projection called the oral arm. Each arm has a ventral ciliated groove leading into the mouth.

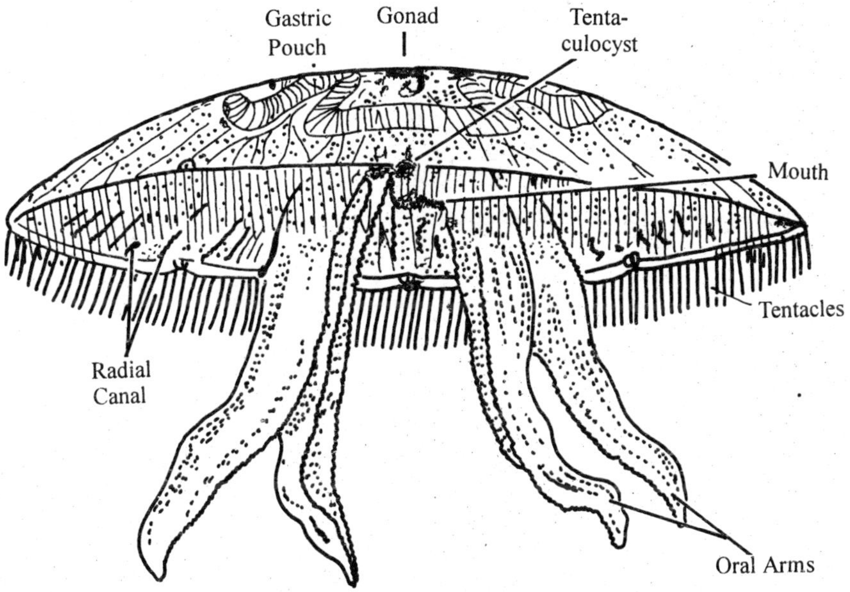

Fig. 18.3 Jelly fish

Systematic Position

Phylum : Coelonterata
Sub Phylum : Cnidaria
Class : Scyphozoa
Type : *Aurelia aurita*

CORAL

1. It is a colonial marine animal commonly found in Mediterranean at a depth of 10-30 fathoms.
2. It is commonly called 'red coral'.
3. The colony is upright and branched and is supported by calcareous axis of fused spicules.
4. The colony is dimorphic as there are two type of zooids (i) nutritive zooids, the autozooids and (ii) for circulation of water in the colony, the siphonozooids.
5. The zooids are white in colour and the colony is dark red due to the presence of red calcareous spicules.
6. Gonads are borne by siphonozooids.
7. Planula develops inside the zooids and so that vivipary is found in Corallium.
8. The Corallium is of great economic importance as the precious red coral of commerce known as red moonga in N.India which is obtained from this animal.

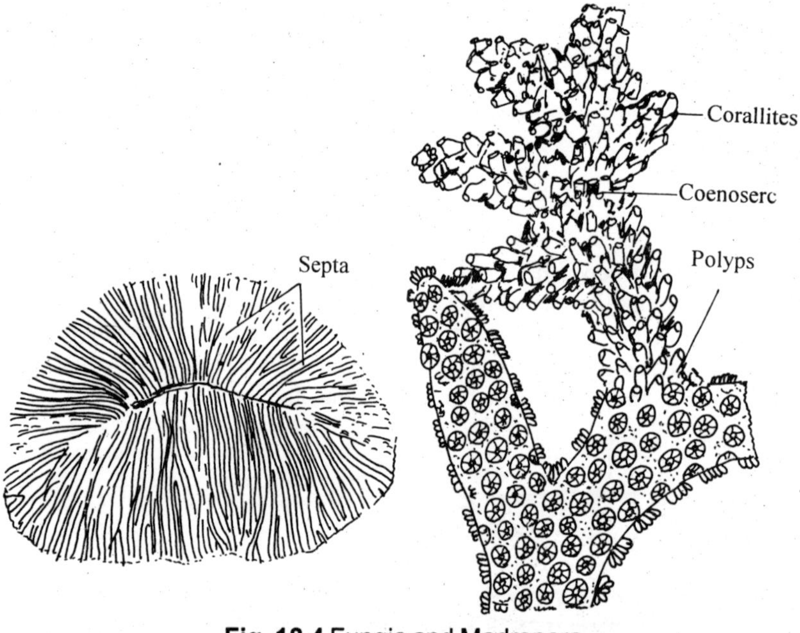

Fig. 18.4 Fungia and Madrepora

Systematic Position

Phylum : Coelenterata
Sub-phylum : Enidaria
Class : Anthozoa
Type : *Fungia and Madrepora*

LIVERFLUKE

Liver fluke is an endoparasite in sheep, cattle, goats etc. It lives in the bile duct of sheep. Rarely it may be even found in the liver of man also. The body liver fluke is flat and unsegmented about 25mm long and 12-13mm broad. The body is reddish brown in colour and bilaterally symmetrical. The anterior region is broad while the posterior region is pointed. The anterior region has the head lobe with an oral sucker enclosing the mouth. On the ventral side is a ventral sucker (acetabulum) helping the attachment of the parasite to the host. Anterior to the ventral sucker is found the gonopore through which eggs pass out.

The body wall of liver fluke is triploblastic and is made up of five layers - cuticle, basement membrane, circular layer of muscles, longitudinal layer of muscles and a layer of diagonal muscles.

Liver flukes are hermaphrodites. The male reproductive system consists of a pair of testes from which arise two tubes (vas deferens) which join to form a seminal vesicle. This opens to the genital atrium through the ejaculatory duct. The female reproductive system has a single, tubular, branched ovary found on the right side of the body in front of the testes. The branches of the ovary join to form an oviduct which enlarges to from the uterus.

The life history of liver fluke is complicated and involves two hosts - pond snail (intermediate host) and sheep (primary host). It also comprises the following larval stages - miracidium, sporocyst, redia, circaria and metacercaria. The miracidium larva is hatched out of the fertilized egg from sheep. It is a free swimming larva, lives for about 8 hours and gains entry into the pond snail. It turns into sporocyst within the liver of the pond snail where another larval stage the redia larve. Finally the cercaria comes out of the body of the pond snail and swims freely in water for about 2-3 hours and settles on the leaf of any water plant. It encloses itself by a gelatinous wall and is now known as the metacercaia. The metacercaria are swallowed by sheep and the life cycle continues.

The infection by liver fluke causes liver rot in sheep and also bilary colic in man.

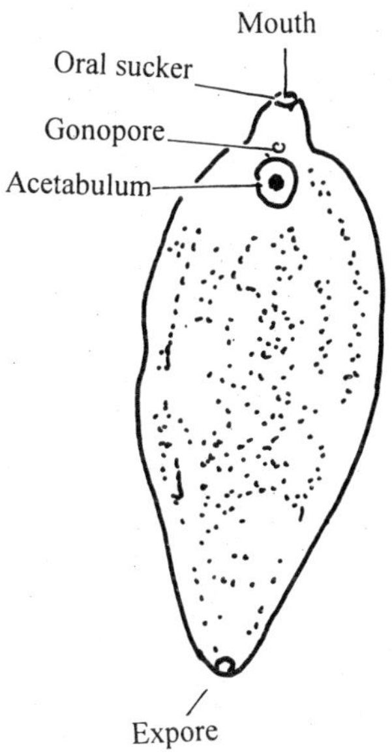

Fig. 18.5 Liverfluke

Systematic Position

Phylum	:	Platyhelminthes
Class	:	Trematoda
Order	:	Digenea
Type	:	*Fasciola hepatica*

TAPE WORM

Taenia solium or pork tape worm is commonly found wherever pork is eaten. It is found as an endoparasite in the gut of vertebrates including man.

The body of tapeworm is elongate and flattened dorsiventrally like a ribbon. The length may be as much as 3-5 metres. The body is divided into three regions scolex, neck and strobila. Scolex or the head region is a pear shaped structure and has a terminal cone - rostellum which is surrounded by two rows of chitinous hooks. There are four suckers a little below the restellum which help in attachment. The neck is situated immediately behind the scolex. It is unsegmented and narrower than the scolex and strobila. The strobila constitutes the main body and is made up of a number of units called proglottids. Each tapeworm may have as many as 800-1000 linear repeating units.

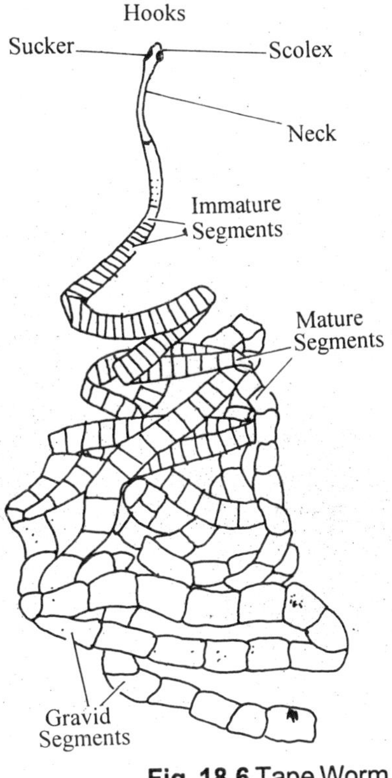

Fig. 18.6 Tape Worm

A mature proglottid consists of a set of male and female reproductive organs.

There are no respiratory organs in tapeworm; respiration taking place through the general body surface. Excretory system is same as found in trematodes - the protonephric type.

The nervous system has to ganglia situated in the scolex region.

Tape worm is hermaphrodite. The male reproductive system consists of numerous minute testes, efferent ducts, vas deferens (a large tube) and cirrus sac. The cirrus sac opens to genital atrium which open to the exterior through the gonopore.

The female reproductive system has a bilobed ovary which leads to oviduct. Uterus arises from the oviduct. The oviduct develops into an enlarged ootype where it joins the vitelline duct. Uterus arises from the oviduct. The vagins is a narrow tube arising from the junction of oviduct with the vitelline duct and opens to exterior through the gonopore.

Systematic Position

Phylum	:	Platyhelminthes
Class	:	Cestoda
Order	:	Taeniodea
Type	:	*Taenia solium*

ANIMAL TAXONOMY

Round worm is the most common endoparasite found in the intestine of man. It is worldwide in its distribution. In human beings *Ascaris* is more common among children than in the adults because of their propensity to contact contaminated material. The round worms usually inhabit the intestine, but in some cases they may even enter the bile and pancreatic ducts. They may also perforate the gut walls causing peritonitis.

Sexes are separate in Ascaris. Both male and females are elongated, cylindrical and have pointed ends. Males are 15-31 cm long and 2-4 mm in diameter while the females are 20-35 cms long and 4-6 mm in diameter. In addition, the males have a curved posterior end with two copulatory spicules protruding from the slit like opening. The females have an uncurved posterior end.

There are four longitudinal lines of streaks running throughout the length of the body. A triangular mouth is present at the anterior end of the body. It is surrounded by three denticulated (teeth like) lips. A circlet of four sensory papillae are found on the lips. There is an excretory pore situated midventrally a little behind the mouth. In females, a gentile pore is present midventrally at about 1/3 distance from the anterior end. The anus is also present only in female.

The body wall is made up of a multi layered cuticle, epidermis and muscle layers. The body cavity is a pseudocoelom filled with a fluid.

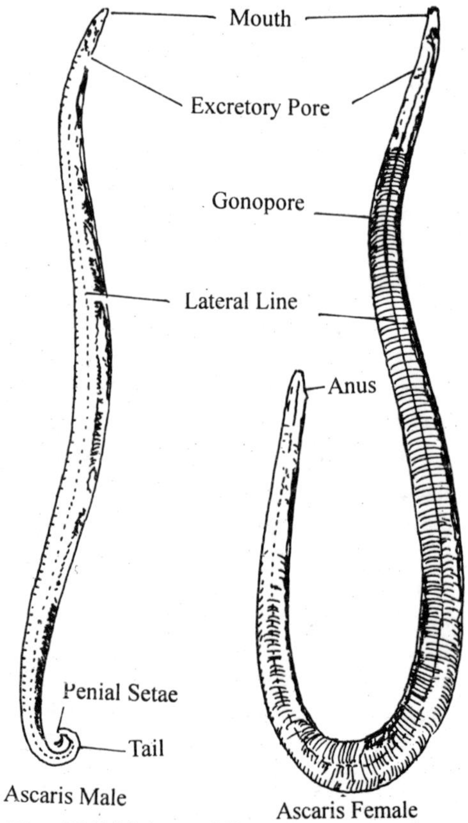

Fig. 18.7 Male and Female round worms

Systematic Position

Phylum : Aschelminthes
Class : Nematoda
Order : Acaroidea
Type : *Ascaris lumbricoides*

LEECH

1. They are marine, aquatic, terrestrial or free living. Some are ectoparasitic and blood-sucking annelids.
2. Body is segmented and is marked externally by ring-like annuli.
3. Presence of anterior and posterior cup-shaped suckers which are the organs of adhesion and locomotion.
4. The ventral mouth lies in the cup-shaped hollow cavity of anterior sucker. The anus is dorsal and situated just above the posterior sucker.
5. The coelom is greatly reduced due to the presence of connective tissue and is represented by haemocoelomic sinuses.
6. The true blood vessels with definite muscular walls present.
7. The nervous system includes a brain united by small oesophageal connectives to a double ganglionated ventral nerve cord.
8. Nephridia are segmentally arranged and open independently to exterior by nephridiopore, which are the organs of excretion.
9. Hermaphrodite, the testes are many and segmentally arranged, whereas ovary is single paired. Eggs are laid in cocoons.
10. Fertilization internal, development direct without larval stage.

Fig. 18.8 Leech

Systematic Position

Phylum : Annelida
Class : Hirudinea
Order : Gnathobdelia
Type : *Hirudo granulosa*

NEREIS

The body of Nereis is elongated, cylindrical and bilaterally symmetrical. The body gradually tapers towards the posterior end. The dorsal surface of the body is convex but the ventral surface is flat. The size varies from few to 40 cms.

The body is divided into approximately 200 or more segments or metameres.

The body is divided into three well defined regions head, trunk, pygidium. Head is the anterior most part of the body. It consists of 2 parts prostomium and peristomium. Prostomium is present before the mouth and is triangular in outline. There are two tentacles in this region. The peristomium is a large ring like round part present behind the prostomium on its ventral side it bears the mouth.

The trunk comprises practically the entire body except for the head and the last segment - pygidium. There are number of segments in the trunk region. Each segment on either lateral side beares a vertical flap like structure called parapodia. The parapodium is two segmented.

The pygidium represents the last segment of the tail. It bears the terminal anus and a pair of long filamentous cirri.

Systematic Position

Phylum	:	Annelida
Class	:	Polychaeta
Order	:	Errantia
Type	:	*Neanthus virens*

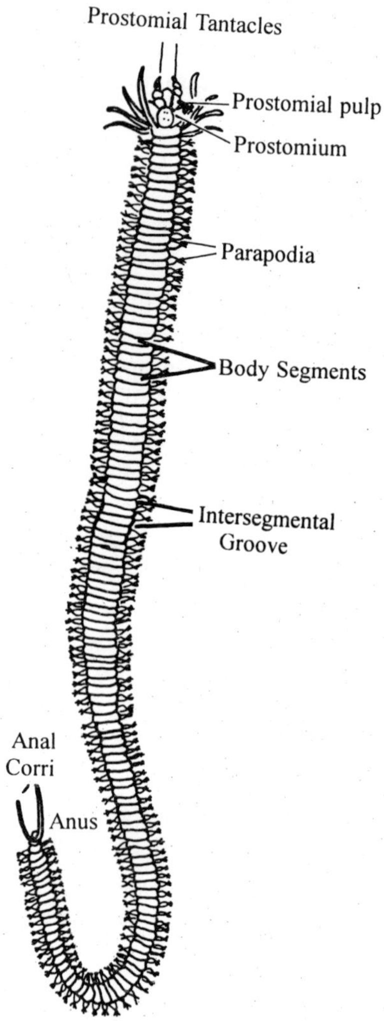

Fig. 18.9 Nereis

PRAWN

Prawn is a bilaterally symmetric animal. The body is elongated and spindle shaped. The size of the adult may measure about 30-75 cms in length. The body is divisible into two distinct regions anterior cephalothorax and posterior abdomen. The cephalothorax is rigid and immovable. It is composed of 13 segments and has 13 pairs of appendages. The cephalothorax is covered by a large chitinious shield comprising of two parts dorsal shield and the sternal plate. The dorsal plate consists of two parts namely the anterior dorsal plate and the posterior carapace. The dorsal plate consists of the rostrum. At the base of the rostrum on either side are attached the compound eyes.

The abdomen consists of six posterior movable segments and the terminal conical structure the telson. Each abdomonal segment is covered dorsally by a convex tergum and ventrally by the stermum. The mouth is present at the anterior end of the cephalothorax. It is surrounded by maxillae.

Systematic Position

Phylum	:	Arthropoda
Class	:	Crustacea
Order	:	Decapoda
Type	:	*Palaemon malcolmsoni*

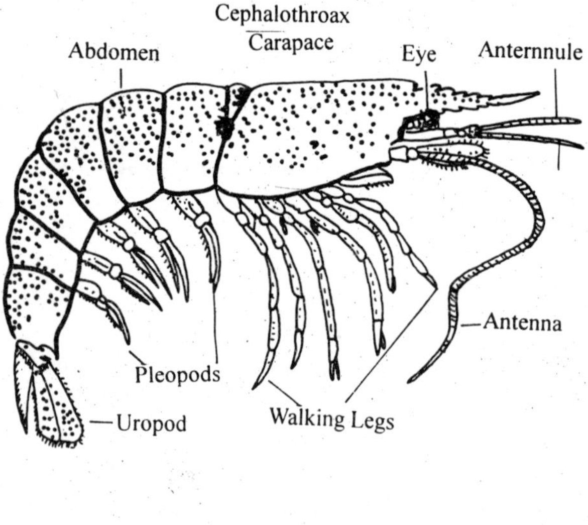

Fig. 18.10 Prawn

MILLIPEDE

1. It is found in damp and dark places under stones and decaying leaves. When disturbed usually roll up into tight coil and emits an unpleasing odour from segmental scent glands. 2. It has cosmopolitan distribution.
3. It is commonly called as 'wire worm'.
4. Body cylindrical and metamerically segmented; differentiated into head and trunk.

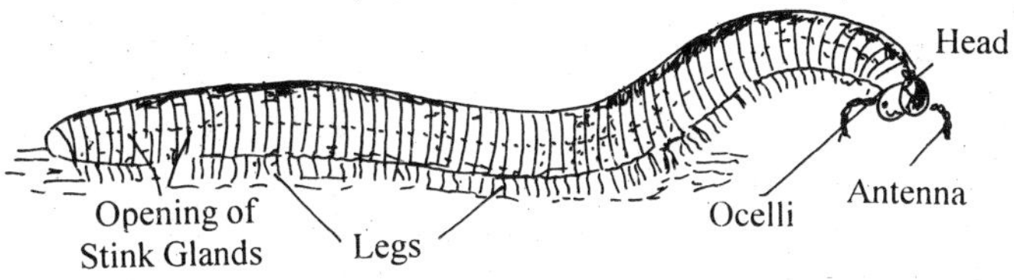

Fig. 18.11 Millipede

5. Head small; antennae short generally ? segment. One pair of maxillae and mandibles forming gnathochilarium.
6. Eyes are minute.
7. Trunk cylindrical consisting of very closely ring like segments from 25 to more than 100 segments.
8. In each segment there are two pairs of legs. Since it has numerous legs it is called millipede.
9. Poison glands and poison claws absent.
10. Vascular system well developed.
11. Respiration by tracheae.
12. Sexes are separate.
13. Genital openings situated at the anterior end of body.
14. In male one or both pairs of legs of 7th segment are modified to form copulatory organs.
15. Herbivorous and even scavanger.

Systematic Position

Phylum	:	Arthropoda
Class	:	Myriapoda
Order	:	Diplopoda
Type	:	Julus sp.

SCORPION

1. These are terrestrial organisms and are found usually hiding under stones, rocks, vegetation etc.
2. They rarely come out of their hideout during day time. They are mostly nocturnal.
3. The body is elongated and has a dorsiventral symmetry.
4. In size scorpions range from 5 - 15 cms.
5. The body is divided into 3 parts namely prosoma, mesosoma and metasoma.
6. The anterior part is called the prosoma and is formed by the fusion of head and thorax.
7. This region is covered by carapace and has a pair of median eyes and 3-5 pairs of lateral eyes.
8. Four of walking legs are also borne on the prosoma.
9. The mouth is present ventrally.
10. There are a pair of large claws called peripalpae.

11. In scorpion the mesosoma and metasoma are called opisthosoma.

12. The mesosomal regions has 7 segments and is covered by a dorsal tergum and a ventral sternum.

13. Ventral side of the mesosoma has 4 pairs of stigmata leading to book lungs internally.

14. The metasoma is the tail like structure and is usually kept raised when the scorpion is walking it has 5 segments and a poision sting at the end.

Systematic Position

Phylum	:	Arthropoda
Class	:	Arachnida
Order	:	Scropinoida
Type	:	Palamnaeus

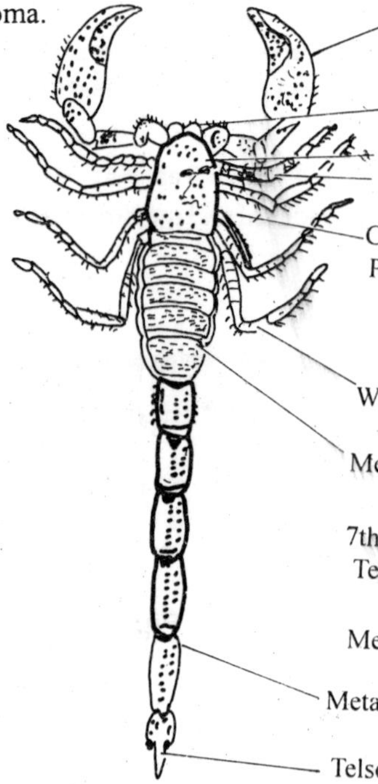

Fig. 18.12 Scorpion

OCTOPUS

1. It is marine and cosmopolitan in distribution generally found on the Atlantic and Pacific coasts.

2. It is commonly called devil fish.

3. Body surrounded by mantle containing chromatophores. They can change their colour according to surrounding medium.

4. Foot modified into eight equal arms surrounding the head and siphon.

5. The arms are beset with suckers which are sessile and without horny rims.

6. Body is covered with tubercles, short, round aborally and without fins.

7. External shell is absent and body with vestiges of an interanl shell.

8. Third arm on the right side in male modified to act interomittent organ or hectocotylized.

9. Like Sepia and Loligo, Octopus is also capable of producing inky liquids which form a smoky screen water.

10. Sexes are separate and there is no metamorphosis as the development is direct. Female cares e eggs.

11. Mouth with sharp and beak-like horny jaws.

Systematic Position

Phylum	:	Mollusca
Class	:	Cephalopodia
Order	:	Octopoda
Type	:	Octopus

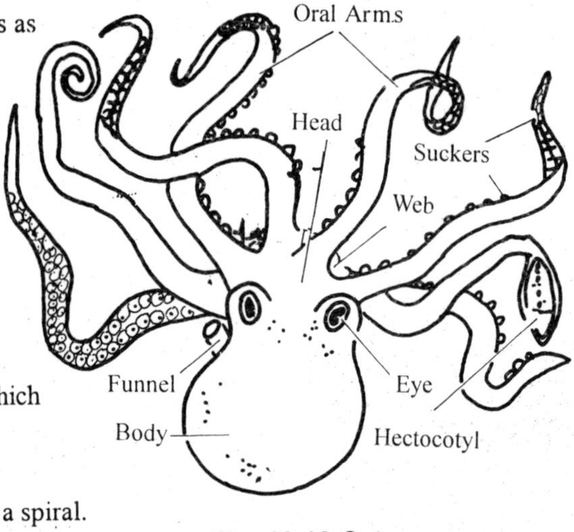

Fig. 18.13 Octopus

PILA (APPLE SNAIL)

1. The body of Pila is enclosed in a calcareous shell which is somewhat conical and coiled round a central axis.

2. The top of the shell is called as the apex.

3. Below this are found successive lines running in a spiral.

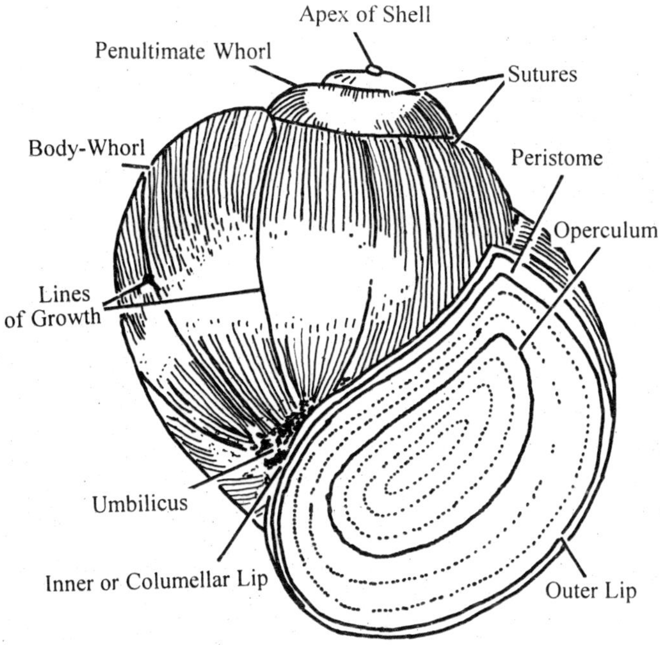

Fig. 18.14 Pila

4. The lowermost spiral ends in an opening - the mouth.
5. The margin of the mouth is called the peristome.
6. The body of the animal is firmly attached to the shell by means of the columellar muscle.
7. The body can be distinguished into foot, head and visceral mask.
8. The head is the anterior part of the body and it over hands the foot.
9. It bears the first pair of tentacles and has a mouth.
10. The foot is large and strongly muscular present in the ventral part of the body. It is roughly triangular in shape.
11. The visceral mask is a spirally coiled hump like structure on the dorsal side and contains all the vital organs.

Systematic Position

Phylum	:	Mollusca
Class	:	Gastropoda
Sub class	:	Prosobranchia
Order	:	Megagastropoda
Type	:	*Pila globosa.*

FRESH WATER MUSSEL

1. It is found in rivers, lakes and ponds.
2. The body is covered by means of a typical bivalve shell.
3. Each valve bears a dorsal projection called umbo.
4. The two valves are attached by elastic protein band called the hinge ligament.
5. When the shell is removed the soft body of the animal is exposed.
6. It is covered by a thin semi transparent fold of skin called the mantle.
7. The mantle also has two folds corresponding to two valves.
8. The mantle cavity is a space present between the two mantles.

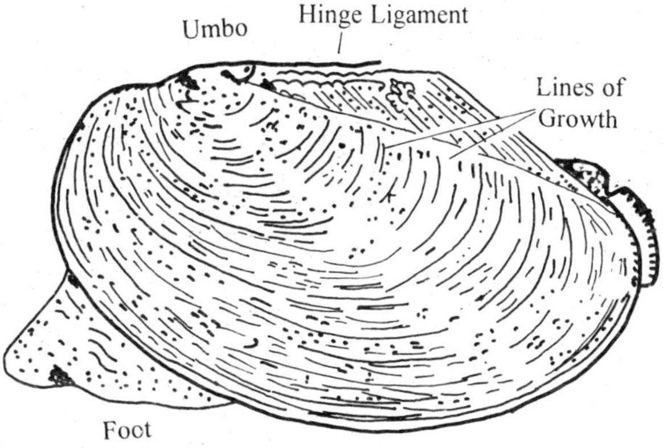

Fig. 18.15 Unio

9. It consists of visceral mask, gills, foot etc.
10. The visceral mask occupies the dorsal part of the mantle cavity.
11. Gills are a pair of long, thin, double folded structures hanging freely into the mantle cavity.
12. The muscular part of the body present on the ventral surface constitutes the foot.
13. In fresh water muscle a distinct is not present.
14. There are no eyes or tentacles. However there is a large mouth in the anterior part of the body.

Systematic Position

Phylum	:	Mollusca
Class	:	Pelecypoda
Order	:	Eulamellibranchiata
Type	:	*Unio mergratifera*

STARFISH (SEA STAR)

1. Body flattened, star shaped with five long slender arms radiating from central disc.
2. Oral and aboral surfaces are distinctly marked.
3. The oral surface contains centrally placed mouth which communicates with 5 broadly opened ambulacral grooves having double rows of tube feet.
4. Tube feet without suckers.
5. The aboral surface has madreporite in one of its interradii.
6. Anus and pedicillariae absent.
7. The stone canal is extremely complicated due to interanl development of ridge.
8. The optic cushion contains retinal cells.
9. Sexes are separate. The development includes brachiolaria larva.

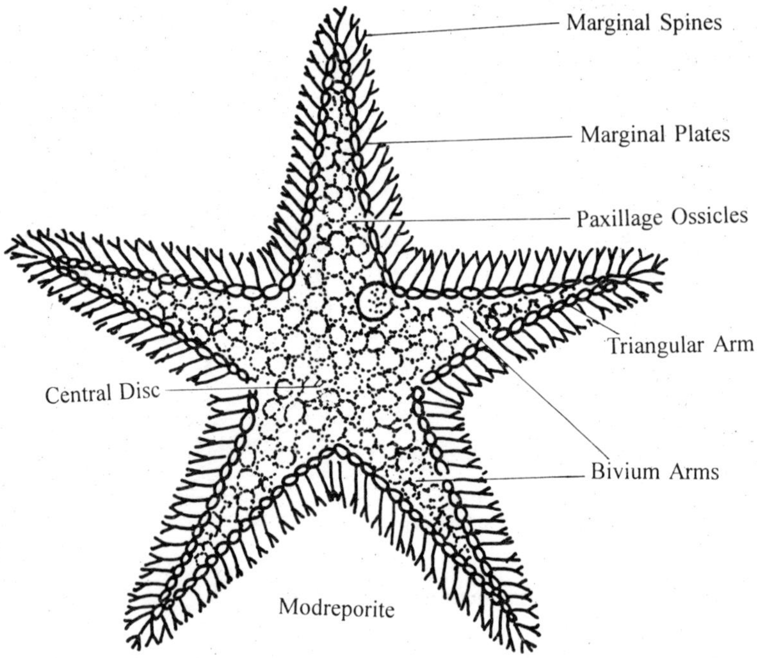

Fig. 18.16 Starfish

Systematic Position

Phylum	:	Echinodermata
Sub - Class	:	Eleutherzoa
Order	:	Asteroidea
Type	:	*Asteropecten*

SEA URCHIN

1. This is usually found on the sea coasts particularly where it is rocky.
2. It is usually found in shallow waters.
3. The body is roughly spherical in outline with a flattened upper and lower poles.
4. The entire body is covered by a rigid shell known as corona.
5. Closely fitting calcareous plates constitute the corona.
6. The entire surface of the corona is covered by spines.
7. The mouth lies in the centre and is surrounded by the peristome.
8. The entire body surface is divided into alternating ambulacral and inter ambulacral areas.
9. Tube feet are found in the ambulacral region.

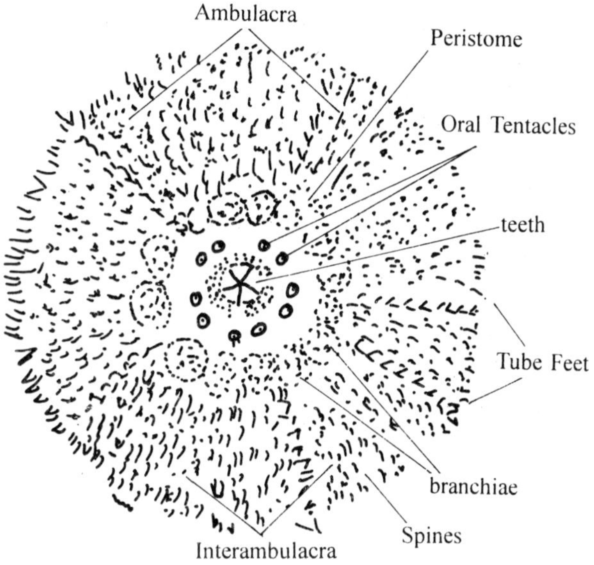

Fig. 18.17 Sea Urchin (Oral view)

Systematic Position

Phylum	:	Echinodermata
Sub phylum	:	Eleutherozoa
Class	:	Echinoidea
Type	:	*Echinus*

CHAPTER 19

CHORDATA

SHARK

1. Shark is a typical example of a marine cartilagenous fish.
2. All the bodily features of the fish are typically adopted to the aquatic life.
3. The body is spindled shaped and streamlined for easy movement in water.
4. There is a lateral line running along the length of the body.
5. The head region is extended into a dorsiventrally flattened snout.
6. The region behind the head is the trunk region.
7. It bears pectral fins, pelvic fins, anal fins and the first and the second dorsal fin.
8. The tail has an extended into a codal fin.
9. The mouth is situated in the ventral position having very sharp teeth.
10. On either side of the snout are present a pair of nostrils.
11. A pair of eyes are also present near this region.
12. On either side of the head are present five pairs of gills.
13. The entire body is covered by plcoid scales.

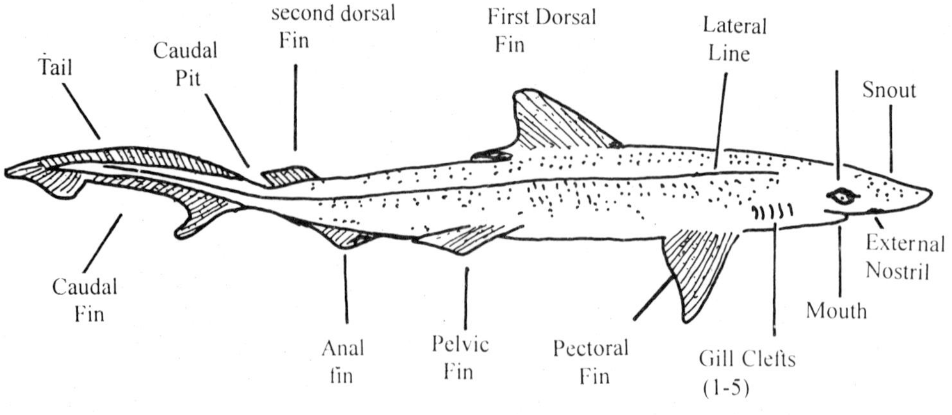

Fig. 19.1 Shark

Systematic Position

Phylum	:	Chrodata
Sub -phylum	:	Craniata
Class	:	Pisces
Sub-class	:	Selachii
Type	:	*Scoliodon sorrokowah*

BONY FISH (Eel)

1. This is a marine organism and has a long snake like body.
2. The entire body is covered by means of minute scales.
3. The mouth is situated in the anterior part it is covered by two lips.
4. The lower lip is slightly longer than the upper lip.
5. On the upper lip is found a pair of nostrils.
6. The eyes are covered by means of scales which are transparent.
7. To help in respiration there is a single gill slit which leads into the gill chamber.
8. There are no pelvic fins but the petral fins are present behind the gills.
9. The dorsal fin is very long and joins the caudal and ventral fin.

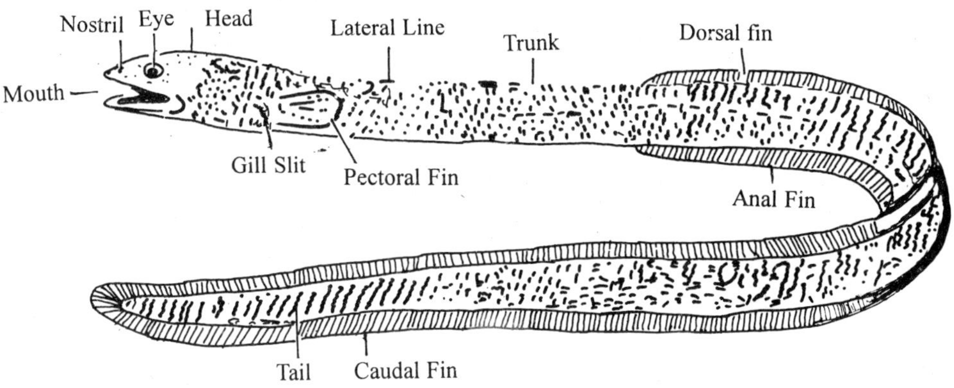

Fig. 19.2 Anguilla

Systematic Position

Phylum	:	Chordata
Sub phylum	:	Vertebrata
Class	:	Pisces
Order	:	Teleostomi
Type	:	*Anguilla* sps.

BONY FISH-2 SEA HORSE

1. This is found in sea waters.
2. The body consists of the head which is produced into a tubular snout.
3. The snout resembles that of a horse. Hence the name sea horse.

4. The top of the snout has a small mouth.

5. The entire body is covered by means of closely packed bony scales.

6. There is also a prehensile tail in the posterior part of the body.

7. Small filament like structures arise from the dorsal surface of the body.

8. A pectral fin is present behind the gill slit.

9. There are no pelvic fins but dorsal fin is present.

10. One unique feature of the sea horse is the presence of a prood pouch in the male for nurtring the eggs.

Systematic Position

Phylum	:	Chordata
Sub phylum	:	Vertebrata
Class	:	Pisces
Order	:	Teleostomi
Type	:	*Hippocampus*

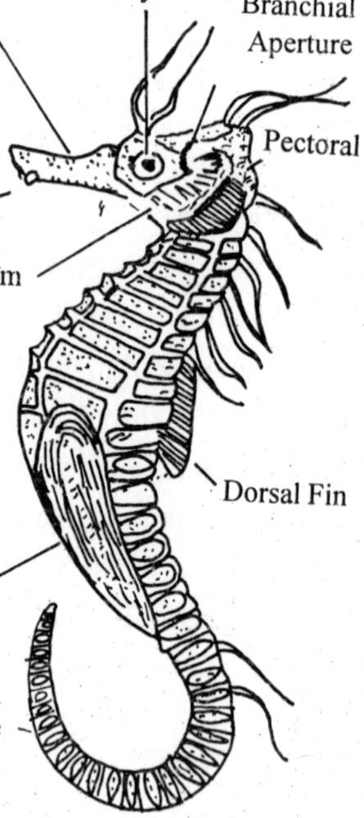

Fig. 19.3 Hippocampus

FROG

1. Frog is a amphibious animal present on the land as well as on water.
2. It has adoptations to suite both the terrestrail and the aquatic life.
3. The body is divided into three parts. Namely head, trunk and limbs.
4. The head is triangular in shape and is extended into a small snout.
5. On either side of the head are present a pair of bulged eyes.
6. The eye is covered by an immovable upper link and movable lower eye lid.
7. The entire body is covered by means of the slime.
8. A pair of nostrils are present at the tip of the snout.
9. A nictitating membrane is present surrounding the eye.
10. It helps the frog while swimming under water.
11. The trunk is broad and somewhat flattened and has a small hump on the dorsal side.
12. Frog has got one pair of hind limbs and one pair of fore limbs.
13. The limbs are pentadactyl.
14. Fore limbs are shorter than the hind limbs and have only four digits.
15. The digits are joined by membraneous webs.

Systematic Position

Phylum	:	Chordata
Sub phylum	:	Vertebrata
Class	:	Amphibia
Type	:	*Rana tigrina*

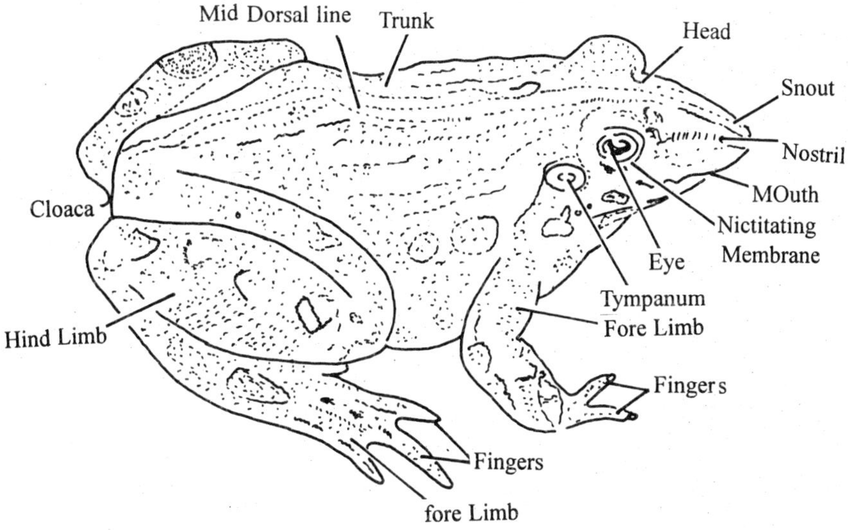

Fig. 19.4 Frog Externals

COBRA

1. It is one of the most poisonous snakes found in our country.
2. The body is very long, cylindrical and has a tapering tail.
3. There is a distinct head followed by the trunk.
4. The entire body is covered by scales.
5. The ventral side of the body has large ventral shields.
6. The neck of the body can be expanded into characteristic hood and it has a typical 'U' shaped marking.

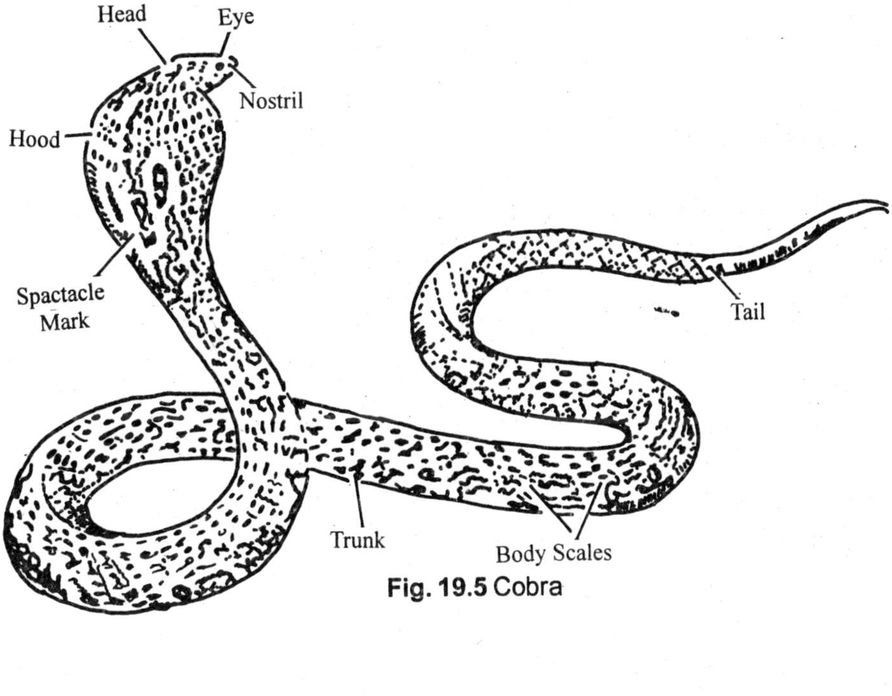

Fig. 19.5 Cobra

7. The mouth is present in the head region, it has a pair of nostrils and eyes.
8. The eyes are immobile and are covered by transparent scales.
9. The tounge is forked and a pair of poison fangs are present.
10. The animals are oviparous.

Systematic Position

Phylum	:	Chordata
Sub phylum	:	Vertebrata
Class	:	Reptilia
Sub class	:	Diapsida
Order	:	Ophidia
Type	:	*Naja naja*

CHAMELEON

1. Chameleons are normally found on trees.
2. They are adopted to the arboreal life.

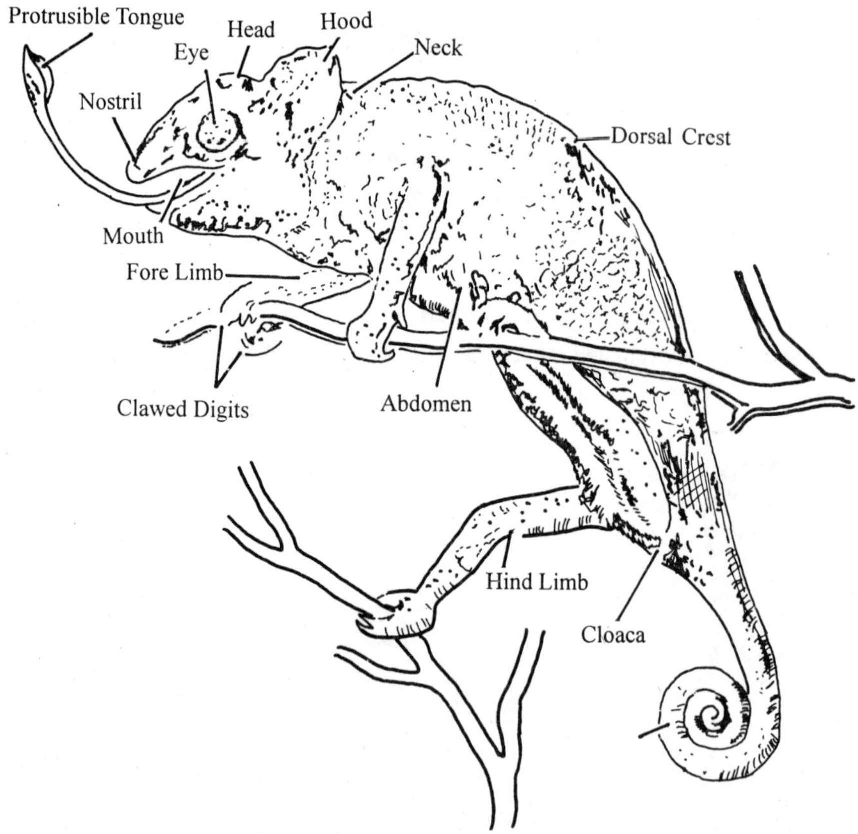

Fig. 19.6 Chameleon

3. The body of the animal has three parts head, trunk and the prehensile tail.
4. The head is triangular in shape and has a projection on the dorsal side.

5. There is a wide mouth enclosing a very long extendable sticky tongue.
6. This tongue helps in the catching of the prey even from some distance .
7. The eyes are mobile and are protruded.
8. The eyes can rotate independently in any direction.
9. The eye lids are fused but there is a small opening to allow for vision.
10. The body is covered by a scally skin.
11. The most remarkable part of the skin is its ability to change the colour according to the surroundings. This helps the animal in mimicry.
12. There are two pairs of limbs hind limb and the fore limb.

Systematic Position

Phylum	:	Chordata
Sub phylum	:	Vertebrata
Class	:	Reptilia
Sub class	:	Diapsida
Order	:	Squamata
Type	:	*Chameleon vulgaris*

TURTLE

1. The turtle is a marine inhabitant.
2. The body is covered by means of a bony shell. It has two parts a dorsal carapace and ventral plastron.
3. The head is triangular in shape and has horny jaws.

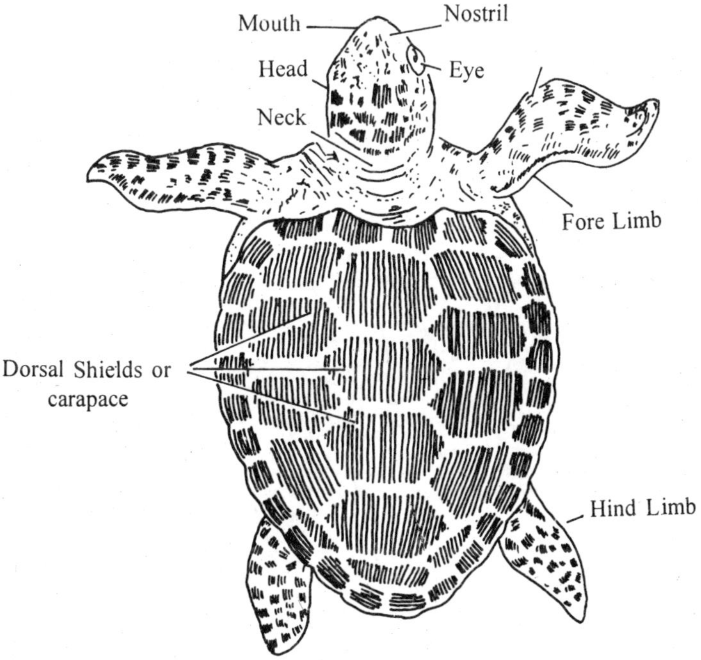

Fig. 19.7 Turtle

4. Unlike in tortoise the cannot be completely retracted into the shell.
5. The body has two pairs of limbs.
6. There are no digits. The limbs function like paddles.
7. There is a very short tail.

Systematic Position

Phylum	:	Chordata
Sub phylum	:	Vertebrata
Class	:	Reptilia
Order	:	Anapsida
Type	:	*Chelone mydas*

CROCODILE

1. Crocodiles are aquatic but they can also move on land.
2. They are usually present in fresh waters like rivers, tanks and ponds.
3. A mature crocodile may be as long as 10-12 feet.
4. The entire body is covered by means of a exoskeleton.
5. This exoskeleton is made up of horny scales.
6. The body is divisible into head, trunk and tail.
7. The head is triangular in shape and is extended into a long snout.
8. Nostrils and eyes are present on the snout in such a way that they will always be above the surface of the water.
9. A stout neck attaches the head to the trunk.

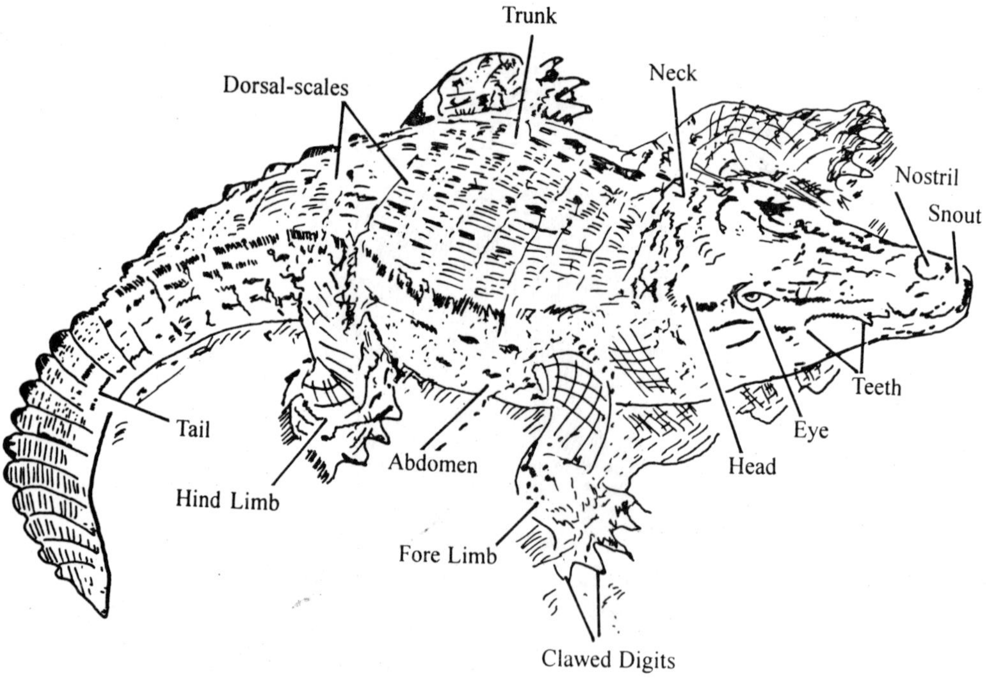

Fig. 19.8 Crocodile

10. The trunk region consists of a pair of limbs.
11. The limbs have clawed digits.
12. The digits are joined by means of webs to help in swimming.
13. The tail part of the body is long and laterally compressed.

Systematic Position

Phylum	:	Chordata
Sub phylum	:	Vertebrata
Class	:	Reptilia
Order	:	Crocodilia
Type	:	*Crocodilus porosus*

PIGEON

1. Pigeon is very common bird found both in rural and urban areas.
2. The body is very light and is adopted to the aerial habit.
3. The body is also streamed lined like a boat to help in air travel.
4. The head region consists of a pair of eyes.
5. The mouth on its either side has two beaks.
6. The eyes are round and have nicutitating membranes in addition to the eyes.
7. The fore limbs are present in the body are modified into wings to help in flying.
8. There are no teeth in the mouth.

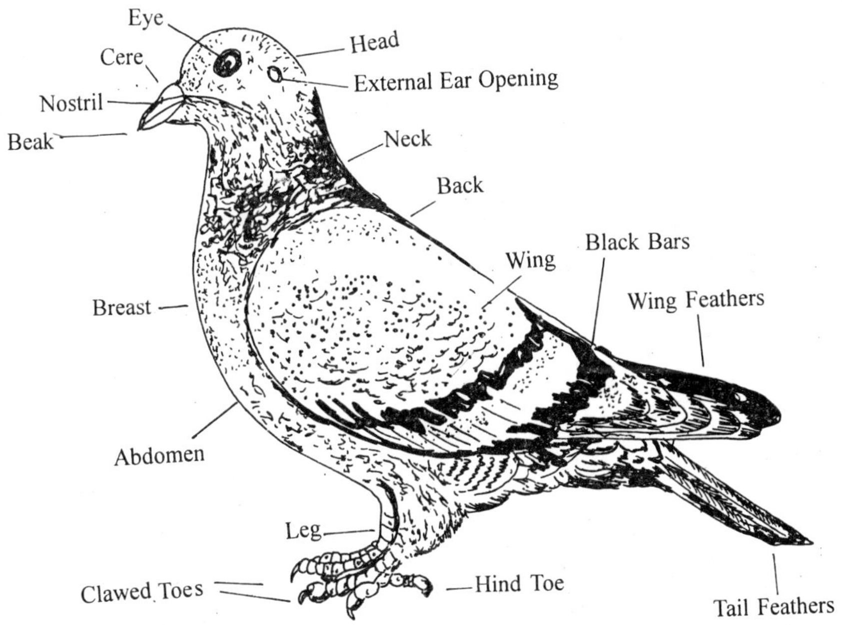

Fig. 19.9 Pigeon

9. The wing is covered by means of quill feathers.
10. In addition to quill feathers body feathers are also present which protect the body.

Systematic Position

Phylum	:	Chordata
Sub phylum	:	Vertebrata
Class	:	Aves
Order	:	Carinatae
Type	:	*Columbia livia* (Pigeon)

DUCK

1. Duck is a domesticated bird and is semi aquatic in its habitat.
2. Even though wings are present they are very weak and the ducks cannot fly.
3. The body of the duck is boat shaped.
4. It has a head and a long neck.
5. The head portion has a long beak or bill.
6. The bill is surrounded by horny epidermis.
7. The beak has thin ridges called strainers which help the bird in filtering muddy water.
8. There are two legs which are short and have a webbed feet.

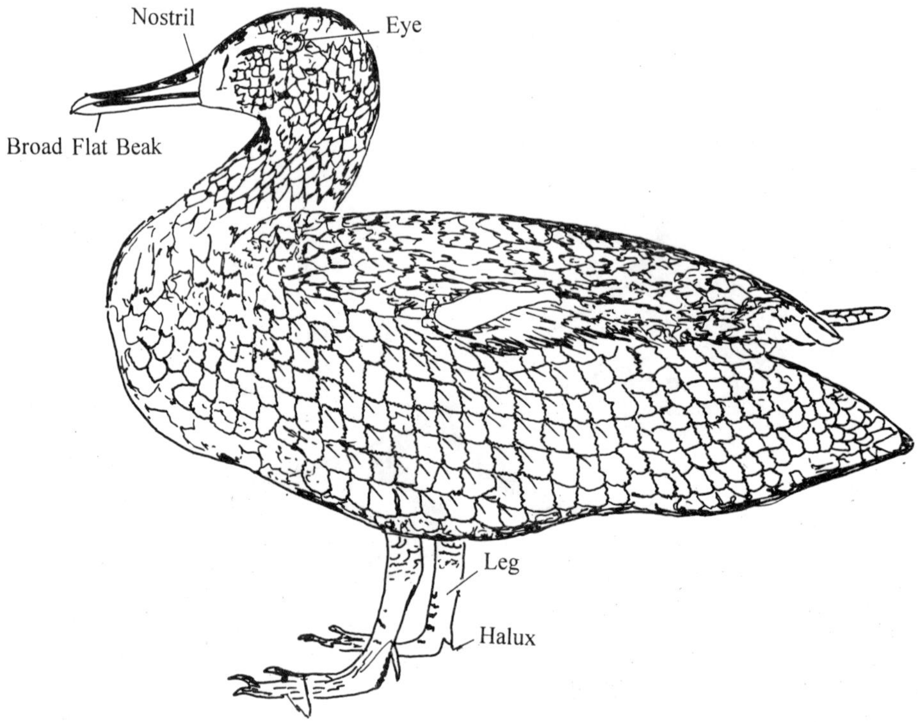

Fig. 19.10 Duck

Systematic Position

Phylum	:	Chordata
Sub phylum	:	Vertebrata
Class	:	Pisces
Sub class	:	Carinatae
Order	:	Anseriformes
Type	:	*Anas* sp. (Duck)

CROW

1. The crow is a very common bird found in urban and rural areas near the human dwellings.
2. It is a medium sized bird with the length ranging between 13-17 inches.
3. The crow is black in colour but the neck is somewhat greyish.
4. There are two legs which have four clawed toes.
5. There are two powerful wings which helps the bird in flying.
6. The beak is hard and helps the bird in tearing the food.
7. The bird is omnivorous and acts as a scavanger.

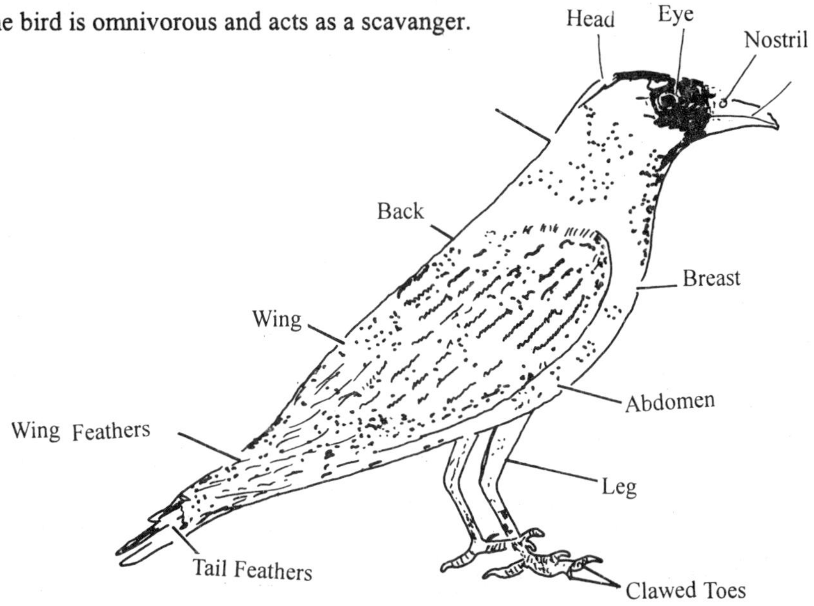

Fig. 19.11 Crow

Systematic Position

Phylum	:	Chordata
Sub phylum	:	Vertebrata
Class	:	Pisces
Order	:	Carinatae
Type	:	*Corvus splendons*

BAT

1. It is a flying mammal.
2. They are usually noctornal.
3. They usually hang down from the branches of trees.
4. The head looks like that of a fox.
5. It has a pair of pinna and a small snout with nostrils.
6. The wing is actually membranous and is called the patagium.
7. There are a pair of limbs.
8. The digits in the hind limbs are clawed in helps the animal in fixing itself to the branches of trees.
9. A remarkable feature of the bat is its ability to send sound waves and receive it back like radar.
10. There is no tail.
11. The body is covered by means of a soft fur.

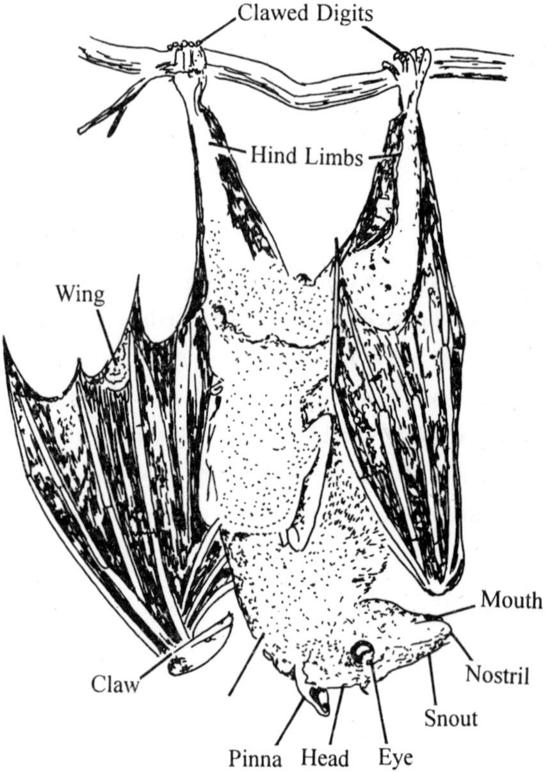

Fig. 19.12 Bat

Systematic Position

Phylum	:	Chordata
Sub phylum	:	Vertebrata
Class	:	Mammalia
Order	:	Chiroptera
Type	:	*Pteropus medias*

RABBIT

1. Rabbit is a terrestrail animal usually found in grass lands.
2. A fully grown animal is a little more than a foot in length.
3. The entire body is covered by a means of fur which is usually reddish brown in colour.
4. The lower surface of the body is somewhat white in colour.
5. The body is divided into head, neck, trunk and tail.
6. Limbs are present in the trunk region.
7. The mouth is present in the head region and is covered by means of two lips of which upper one is cleft like.
8. The upper lip also has sensory viscus.
9. There are a pair of large external pinnae.
10. There are also a pair of eyes covered by means of lids as well as nictitating membrane.
11. The neck attaches the head to the trunk.
12. There are two pairs of limbs, the fore limbs are shorter and have clawed digits.
13. The tail is short and bushy.

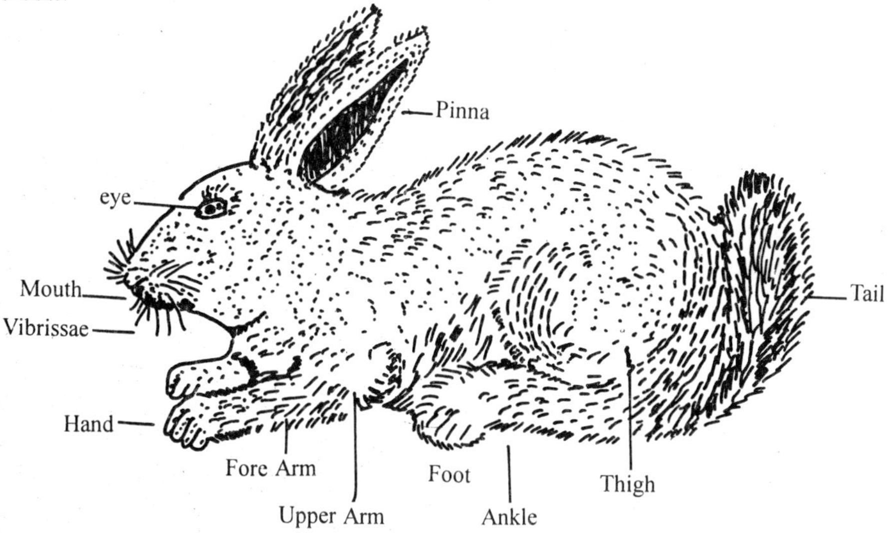

Fig. 19.13 Rabbit

Systematic Position

Phylum	:	Chordata
Class	:	Mammalia
Order	:	Lagomorpha
Type	:	*Oryctologus cuniculus*

STUDY OF COCKROACH

In order to dissect out the digestive system place the cockroach on the dissection board and remove the wings and legs. Cut the body wall along both the sides and lift the body wall and pin it. Release the digestive system and study the various parts.

The digestive system consists of three parts foregut or stomodeum, midgut or mesenteron and hingut or proctodeum.

The foregut consists of mouth, pharynx, oesophagus, crop and gizzard.

The midgut follows the gizzard. Hepatic caeca are present at the junction of the midgut and the gizzard. These hepatic caeca secrete digestive juices. The hind gut is the last part of the digestive canal. It consists of three part namely ileum which is a short tubular portion, a long coiled colon and this is followed by a broad rectum. The rectum opens by means of anus.

Mouth parts

The mouth parts of cockroach consists of the upper lip or labrum, maxillae, mandible, labium and the tongue.

The upper lip is also known as labrum and it is a plate like structure forming the roof of the preoral cavity.

The maxillae are found behing the mandibles, there are three parts in each maxillae, these are the basal segment, the outer maxillary palp and the inner part.

The mandibles are also called the jaws, they are present on either side of the mouth, they have serrated margins and help in cutting the food material.

The labium or the lower lip forms the basal region of the oral cavity. It has three parts namely submentum, middle mentum and pre mentum.

The tongue also known as the hypopharynx is attached to the posterior wall of the preoral cavity.

The frog to be dissected has to be anasthetised with suitable quantity of chloroform. Place the frog on the dissection board and nail it properly. Holding the skin with the help of forceps, cut the skin to expose the internal parts. The abdominal vein can be seen as a bluish line. Cut the body wall along this line towards the anterior end. Cut the pectoral gurdle to expose the internal organs.

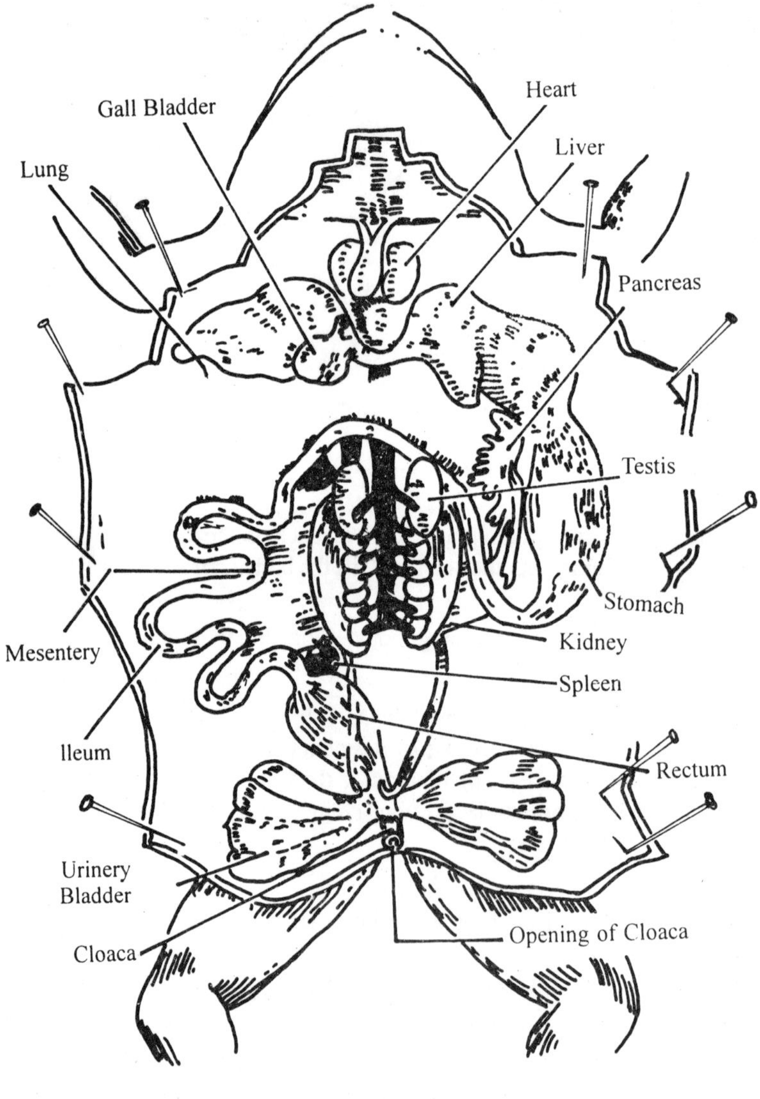

Fig. 19.14 Digestive System of Frog

The digestive system consists of mouth, pharynx, oesophagus, stomach, ileum, colon, rectum, liver and pancreas.

After observing the digestive system remove it and study the urino-genital system. The male urino-genital system consists of yellow colour testes attached to the kidneys. The ureters which arises from the edge of the kidneys join the cloaca. The female urino-genital consists of a pair of large ovries, kidneys and the ureters. On either side of the kidney coiled oviducts are visible.

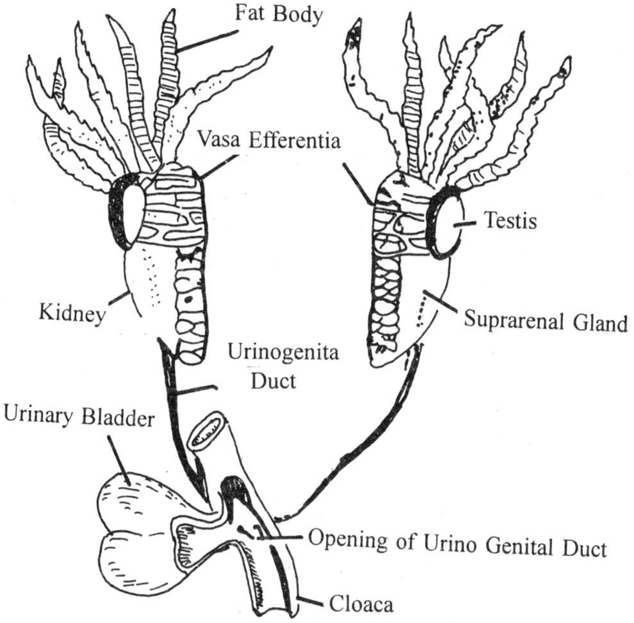

Fig. 19.15 Male Urinogenital System of Frog

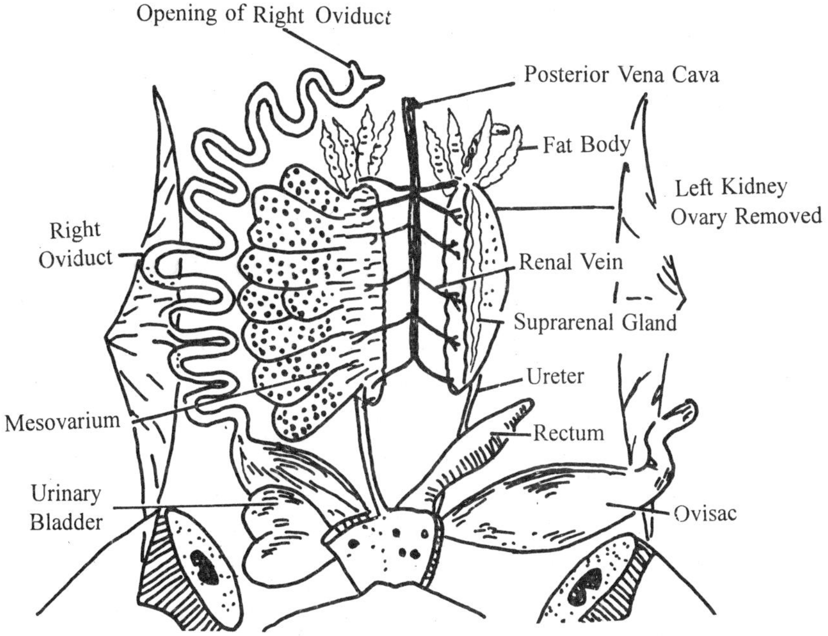

Fig. 19.16 Female Urinogenital System of Frog

CHAPTER 20

ANIMAL TISSUES

STUDY OF TISSUES

I. EPITHELIAL TISSUE

An epithelial tissue may be defined as a tissue composed of one are more layers of cells covering the external or internal surface of the body or may be embryologicaly derived from such a surface.

Structure: Epithelial tissue has cells of various shapes which are closely connected leaving no intercellular space; these form continous sheets. Usually the epithelial cells rest on a delicate non-cellular gelatinous basement membrane or basal lamina. The basement membrane is made up of protein firbres matted together in a polysaccharide ground substance. The basement membrane is secreted by underlying connective tissue. In order to maintain the integrety of the epithelial layer the plasma membranes of adjacent cells are held together by means of intercelluar structures such as desmosomes, inter digitions, tight-junctions etc. In some places the adjacent cells form gap junctions for intercellualr communications and exchange of chemcials.

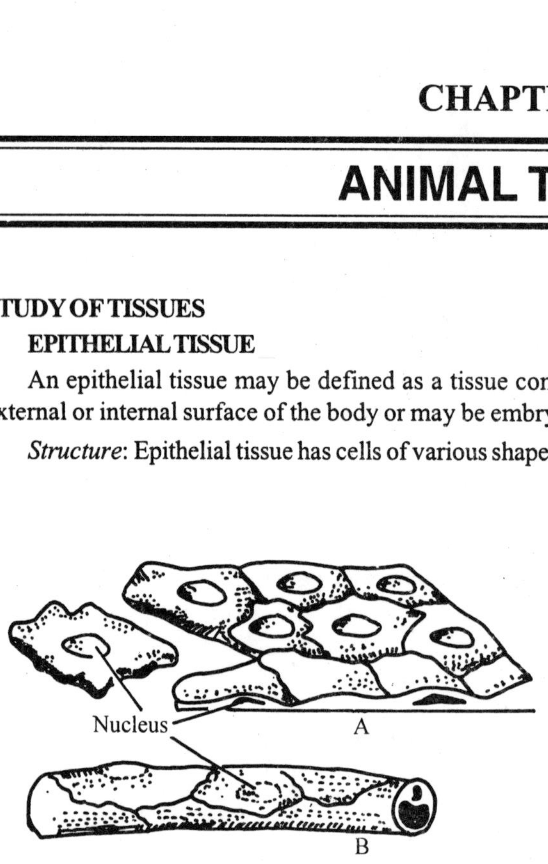

Fig. 20.1 (A) Simple squamous epithelium
(B) Endothelium of the wall of blood

The cells of the epithelium tissue are not provided with blood vessels but still they possess a remarkable capacity of repair and regeneration. After an injury the epithelial cells receive useful materials and excrete waste products to the blood flowing in the blood vessels of the connective tissue present near the basement membrane.

General features of the epithelial tissue:

1. They cover the surfaces and line the cavities and as such they protect other tissues from mechanical injury, drying up, infection and harmful chemcials.

2. They form many useful exoskeleton structures like scales, hairs, feathers, claws, nails, etc.
3. They have secretory activity; many epithelia produce glands and secrete juices like mucous, gastric juice etc.
4. Epithelia that line the uriniferous tubules help in the elimination of the nitrogenous waste materials as urine.
5. Epithelia lining the intestine help in the absorption of digested food.
6. They also help in exchange of gases in the respiratory tract. They bring about exchange of carbon di-oxide and oxygen between blood and air.
7. Many epithelia form a part of the sense organs as in nasal chambers.
8. Epithelia also help in movement of fluids and mucous in the organs where they are present.
9. They also take an important part in the production. Germinal epithelia help in the production of eggs and sperms.
10. Epithelial cells with pigmentation are responsible for the darkness of the cavity of the eye ball.

SIMPLE EPITHELIUM : These are made up of single layer of cells resting on the basement membrane. Usually they cover moist surfaces where there is no wear and tear by friction. Based on their form and function five types of simple epithelia are recognised. These are- Simple Squamous Epithelium, Simple Columnar Epithelium, Simple Cuboidal epithelium, Simple Ciliated Epithelium and Simple Glandular Epithelium.

Simple Squamous Epithelium

This is made up of flat disk like polygonal cells that are closely packed leaving no intercellular spaces. They appear like the arrangement of the tiles on the floor hence the epithilum is also known as the pavement epithilum. The cells of the tissue are connected by fine protoplasmic bridges and are slightly thicker at the centre. The nucleus is slightly flattened and is located at the centre of the cell.

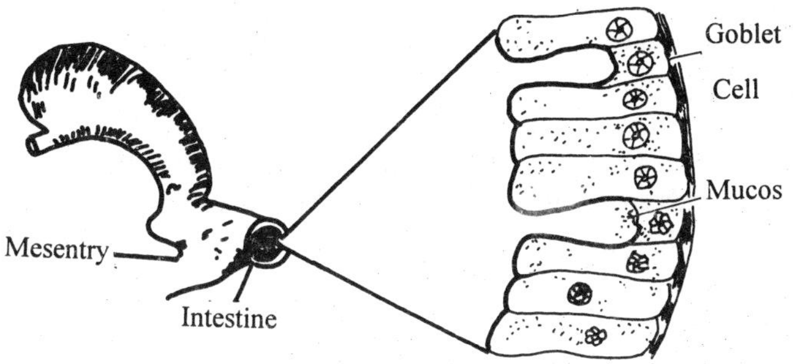

Fig 20.2 Simple columnar epithelium

Simple Cubodial Epithelium

This tissue consists of cells that are usually squarish with a polygonal outline. In a surface view they appear as sheet of cells under the microscope. The cells have single nucleus located in the centre. The cubiodal epithelial cells are located in thyroid follicles, small salivary and pancreatic ducts, lining of the inner ear and proximal parts of the urinary tubules etc. Cubodial epithelium is also found in sweat glands, pigmented layer of retina and in the duct.

Simple Columunar Epithelium

As the very name indicates this epithelium consists of columnar cells that are longer (taller) than broad. The tissue is made up of a single layer of cells. The outlines of the cells appear polygonal in a surface view. Each cell has a flattened and an oval end. There is a single nucleus in each cell which is elongated and usually located either in the centre or a little away from the centre. Usually some mucous secreting cells called *Goblet cells* are found in between the columnar epithelial cells that line the stomach and intestine. The goblet cells have vacoule filled with mucous. The epithelial tissue containing the mucous secreting cells along with the underlying supporting connective tissue is usually refered to as the *mucous membrane* or mucosa.

Simple Ciliated Epithelium

In this tissue the epithelial cells bear at their free ends thin elongated cytoplasmic processes called Cilia (Sing. *cilium*). Each cilium arises from a minute particle called basal granule or blepharoplast that lies internal to the cell membrane. In between the ciliated epithelial cells are present mucous secreting goblet cells. The mucous spreads over the epithelium as a thin coating. The cilia have a beating action (move like the lashes of a whip) as a result of which the mucous and other substances are transfered over the epithelium from one place to another . The cilia also help in shifting small solid particles entangled in the mucous.

The ciliated epithelium is classified into two types- *cuboidal ciliated epithelium* and *columnar ciliated epithelium.*

The Cuboidal ciliated epithelium consists of cuboidal cells with cilia on their free surface. These are found in the urinary tubules of the kidney. The cilia by their lashing action keep the urine moving. The columnar ciliated epithelium has columnar cells with cilia. They are found in the fallopian tubes (oviducts), termianl bronchioles and central canal of the spinal cord of the embryo.

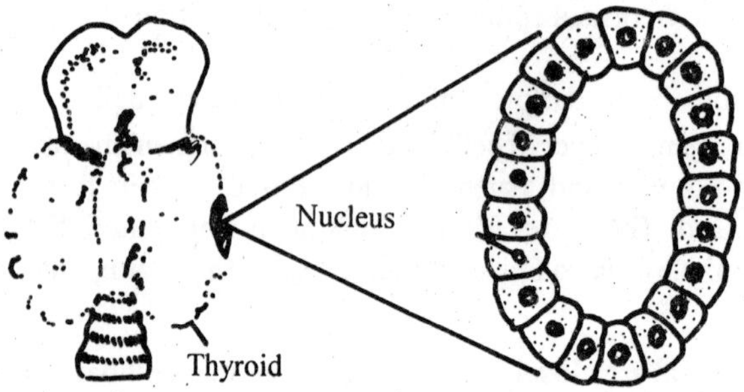

Fig 20.3 Cuboidal epithelium from thyroid gland

Glandular Epithelium: The Glandular epithelium consists of cells which are either cuboidal or columnar. These are specialised for the secretion of chemical substances like mucous or enzymes or hormones. The Epithelial cells fold in different ways to form the glands.

Glands are classified into two basic types. These are *unicellular glands* and *multicellular glands.* Unicellular glands are isolated cells and are distributed between the columnlar cells. These are also called the goblet cells and secrete mucous. Unicellular glands are found in the epithelium lining the intestine.

Multicellular glands are made up of a number of cells. These are formed by the sinking of the glandular epithelial cells into the underlying connective tissue. Multicellular glands are further classified into two types viz., *tubular* and *saccular.*

Tubular: The tubular glands are tube like in appearence and are of two types - *Simple tubular glands* and *compound tubular glands.* The simple tubular glands may be straight coiled or branched. Their secretory parts open into a single duct. Eg. sweat glands in the skin and glandular epithelium in the intestine.

Saccular glands: (Alveolar glands) These glands are flask shaped in appearance and have a wide spherical secretory part called the acinus or alveolus at the inner end of a tubular duct. These glands may also be classified into simple and compound.

The simple alveolar glands may be branched or unbranched. They are found in the cutaneous glands of frog.

The compound alveolar glands are highly branched and consist of several lobules each having many acini. Each acinus opens into the lobule by a short ductlet which ultimately leads to the main duct. Examples of compound alveolar glands are the milk glands of the human beings.

MUSCULAR TISSUE

This is one of the special tissues that possess the unique property of contractility. The tissues are highly elastic and can contract and expand depending on the need. In the living organisms the protoplasm of the cells also possess the property of the contractility but it is more pronounced in muscular tissues. The contraction and expansion of the muscle tissues is in a definite direction ie., along their long axis.

The muscle tissue is made up of elongated thin cells called *muscle fibres* and they are held together by areolar tissue. There is no intercellular tissue between the muscle cells and there is also no matrix. The cytoplasm of muscle fibres is called *sarcoplasm*. Embedded in the cytoplasm are a number of long, fine, parallely arranged protein fibres called *Myofibrils*. These are arranged along the long axis of the fiber. The spaces between the myofirbrils are occupied by a large number of mitochondria (sarcosomes) and glycogen granules. These granules provide the energy necessary for contraction. In some instances the muscle fibres may be bounded by a special membrane called *sarcolemma*. The muscle fibres may have one to many nuclei.

Classification of muscle tissues: Muscle tissues are classified into three categories-

1 Smooth muscle (involuntary muscle)
2. Striated muscle (voluntary muscle)
3. Cardiac muscle

1. *Smooth muscle* : These muscles are also called non-striated or non straight muscles. The muscle fibres are made up of long narrow spindle shaped cells that are generally shorter than striated muscle cells. They average a length of about 0.2 mm. Each muscle fibre has a single nucleus in the central thick portion. The sarcoplasm consists of a number of parallely arranged myofibrils. These are made up of myosin. The muscle cells are not bound by a true sarcolemma; the usual plasma membrane however is present. The cytoplasm consists of mitochondria and endoplasmic reticulum but they are not as many in number as in the striated muscles.

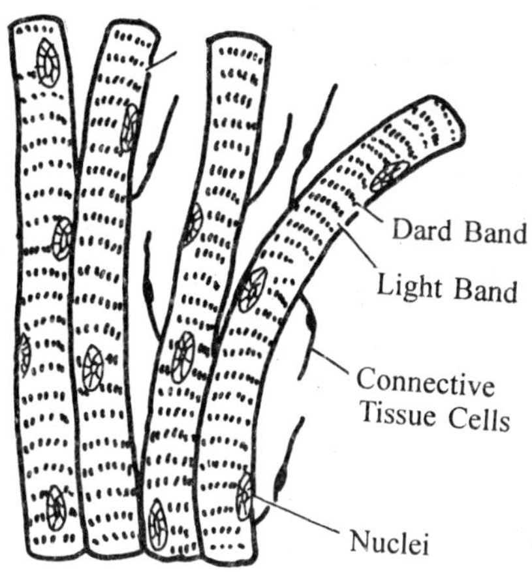

Fig 20.4 Skeletal muscles

2. *Striated muscles* : Also called striped muscles or voluntary muscles or skeletal muscles they constitute about 80% or more of the soft tissues present in the body. The stratied muscle fibres are long, cylindrical, unbranched and have blunt ends. Each fiber may be as long as 40mm. Each muscle fibre is bounded by a conspicuous sarcolemma. The sarcoplasm consists of many elongated, flattened nuclei. In addition there are a number of glycogen granules and mitochondria.

The most striking feature of the striated muscle fibre is the presence of alternating dark and light transverse bands called stripes or striations. Hence the name striated muscles. The dark bands are called anisotropic or A bands. Each A band has at its middle a somewhat light zone called the H zone or Henson's line. The lighter band is isotropic and is called the I band. Each I band has at its centre a dark line called the *membrane of Krause* or Z line or Z band. The dark and light bands alternate regularly. The portion of the myofibril between two Z lines is called a Sarcomere and it constitutes a contractile unit. Thus each sarcomere consists of A band and half of each adjacent I band. Electron microscopic studies reveal the sarcomere is a bundle of thin myofilaments of two types- primary and secondary.

3. ***Cardiac Muscles*** **:** The cardiac muscles are confined to the walls of heart only. They have a unique function of contracting rhythmically and continuously. They generate a unique wave excitation as they pass from one fibre to another.

The cardiac muscle fibres are made up of short cylindrical fibres joined end to end. These are interconnected by crosslinks forming a network. It is this arrangement that helps in the rhythmic contraction so characteristic of the heart muscles.

The muscle fibres are uninucleate and do not have a sarcolemma. The cardiac muscle fibres are striated but the bands are much paler than those of the striated muscle fibres . The cytoplasm consists of a large number of mitochondira and numerous glycogen granules. There are some speicalised regions of the membrane at the end of the fibres which bring about cris-cross junction - these areas are called *intercalated discs* (rather than Z lines of the striated muscles). These intercalated discs help as boosters of the wave action permitting the motion to be transferred from one fibre to the other. The cardiac muscles like skeletal muscles (striated muscles) contract quickly but do not experience fatigue.

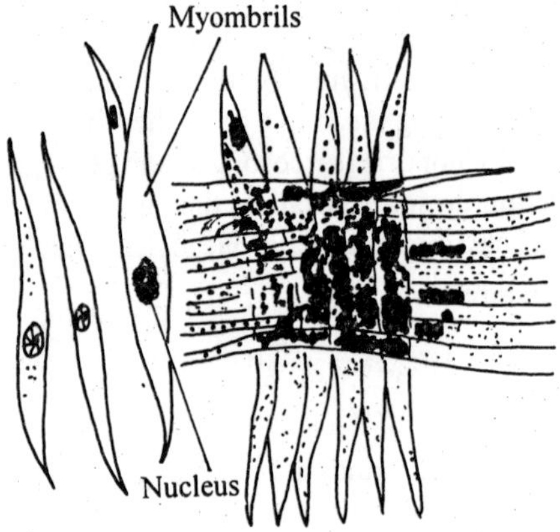

Fig 20.5 Smooth muscles

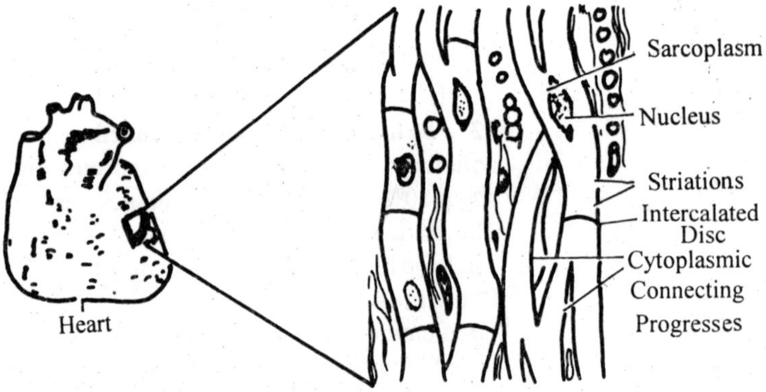

Fig. 20.6 Cardiac muscles

CONNECTIVE TISSUE

Connective tissue is of mesodermal origin and develops from mesenchyma. Structurally the connective tissue consists of various shapes of cells lying in a large amount of non living intercellular or extracellular material called the matrix. As a contrast to the epithelial tissue connective tissues have large intercellular spaces. The matrix or the intercellular material is secreted by the cells themselves. The connective tissue roughly constitutes 20% of the tissues in the body.

In the connective tissue the bulk of the tissue is made up of intercellular substance produced by the cells. This intercellular material consists of fibres and large amount of amorphous, transparent ground substance. Chemically the intercellular matrix is made up of acidic mucopolysaccharides bound with proteins to form hyaluranic acid. In the connective tissue several types of cells are found. They may be freely wandering in the matrix or fixed to a position.

Classification of connective tissue:

Connective tissues

Connective tissue proper	**Skeletal tissues**	**Fluid tissues**
1. Areolar tissue	1. Cartilage	1. Blood
2. Fibrous tissue	2. Bone	2. Lymph
3. Adipose tissue		

Connective tissues are basically divided into three categories viz., *connective tissue proper, Skeletal tissue* and *Fluid tissue* (Vascular tissue)

The cells of the connective tissue proper are immersed in a soft jelly like matrix chemically composed of carbohydrates linked to proteins. The connective tissues proper connect one tissue to another and form protective sheaths around organs.

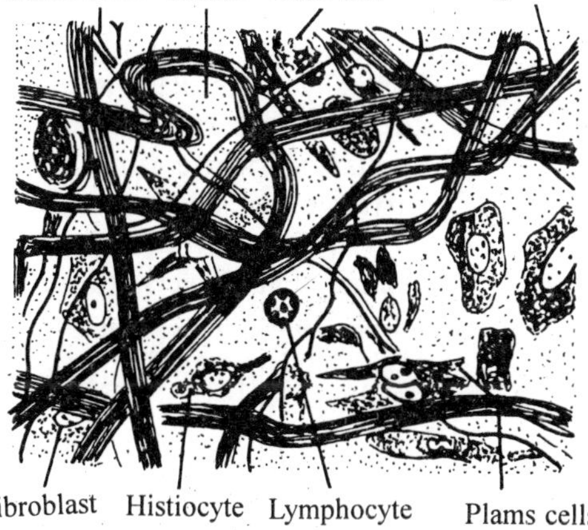

Fig 20.7 Aerolar tissue

Connective tissue proper is divided into five categories namely *Areolar tissue, Adipose tissue, white fibrous tissue, Yellow elastic tissue* and *reticular tissue*. In some cases only three are recognized namely Areolar tissue, fibrous tissue and adipose tissue.

Areolar tissue: This is the most widely distributed connective tissue in the body of the animals. The main function of this tissue is connective. Structurally the areolar tissue has a large amount of transparent sticky matrix enclosing numerous fibrous cells and large quantity of mucin. The fibres in the areolar tissue are of two types *white* and *yellow*. The white fibres occur in fasciles (bundles) and are very fine and wavy. They are bound together by mucin. They are nonelastic and unbranched and are formed of a protein called *collagen*. The collagen fibres are composed of numerous microfibrils which provide tenacity to the tissues.

The Yellow fibres are less in number than the white fibres. They are also much thicker and occur singly. The yellow fibres are branched and elastic with the branches getting interwoven to form a network. Chemically they are composed of a protein called *elastin*. The spaces left between the fibres are called areolae hence the name areolar tissue.

The cells found in the matrix are of three kinds namely *fibroblasts* or *fibrocytes*, *macrophages* or *histocytes* and *mast cells*.

The fibroblasts are elongated flat cells with oval nuclei. They secret the material from which fibres are formed. Those secreting the white fibres lie along with them at the initial stages and later move away. Those producing Yellow fibres are never adjacant to the fibres.

The macrophages are large amoeboid cells with ovoid nuclei. They freely wander in the matrix and engulf foreign particles, microbes, damaged cells etc.

Mast cells are oval shaped cells or irregular shaped cells and have abundant dense granules in cytoplasm It is these cells that produce the matrix.

SKELETAL TISSUES

The function of Skeletal tissue is mainly supportive. The tissues form rigid frame work offering support to the body and provide an infrastructure for the attachment of muscles and bring about locomotion and co-ordination with the muscles. The skeleton found in the higher animals is refered to as endoskeleton as it is found internally.

Skeletal tissues are mainly of two types. Cartilage and Bone.

(1) *Cartilage* : Cartilage is a form of connective tissue which is solid and semirigid. It has a very firm intercellular matrix called the chondrin. Chemically the chondrin is made up of protein chains connected to long chains of di-saccharide hyaluronic acid. Embedded in the matrix are found a number of cartilage cells called chondroblasts or chondrocytes. It is these chondrocytes which secret the matrix. The chondrocytes are somewhat oval and have large nuclei. The cells may remain in groups and are located in a fluid filled space called lacuna. Externally the cartilage is bound by a stiff sheath called perichondrium and it consists of white fibrous tissues. Cartilages lack blood vessels and nerves. The cartilages grow by the addition of new layers of matrix to the outside. Injury to the cartilage generally takes a longer time to heal as there is no blood supply.

Based on the type of matrix and its structure cartilages are classified into three types viz., hyaline, fibrous and calcified.

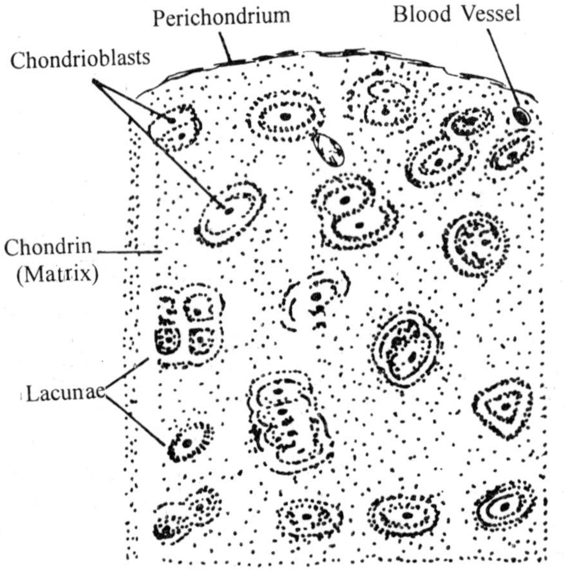

Fig 20.8 T.S. Hyalineo cartilage

Hyaline Cartilage: This is the most common cartilage. The matrix is a clear homogenous translucent substance. Very often a net work of collagen fibres are dispersed in the matrix. The interweaving of the collagen fibres is very fine and very close and often difficult to observe. The name hyaline is given to this cartilage because of a translucent hyaline appearance of the matrix. The matrix consists of many lacunae and a number of chondrocytes. The cartilage is covered by a membrane called perichondrium having blood vessels. It is from these blood vessels that nutritive substances difuse into cartilage. In the body, hyaline cartilage is found in trachea , Bronchial rings, Larynx, Nasal septum, Sternal ribs etc.

BONE

This is a solid rigid connective tissue. The greater part of the skeletal system in most of the vertebrates is made of bone. Like other connective tissues the bone is also basically made up of matrix and bone cells. The bones may be solid when they are small or they may be having a cavity as in the case of long bones such as humerus and femur. The substances of the bone are distinguishable into three regions surrounding the central cavity. These three regions are periosteum, matrix and endosteum.

The central cavity is usually refered to as the marrow cavity. It consists of the bone marrow. The bone marrow stores the reserve food material and also is the place from where new blood cells are produced.

External to the marrow cavity is the endosteum. It consists of white fibrous tissue and contains the bone forming cells which produce new bone material.

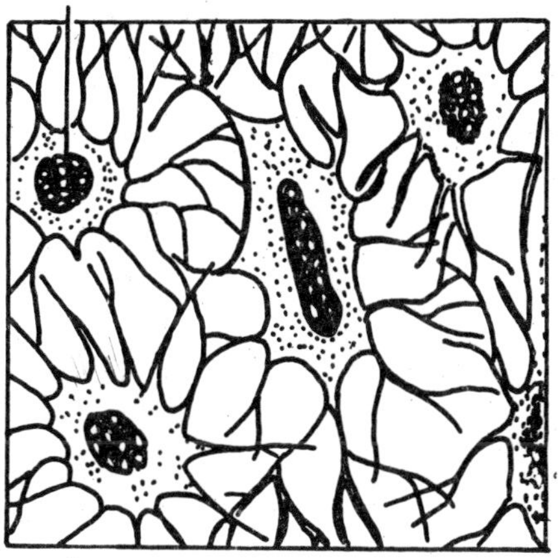

Fig 20.9 Bone cells and their processes

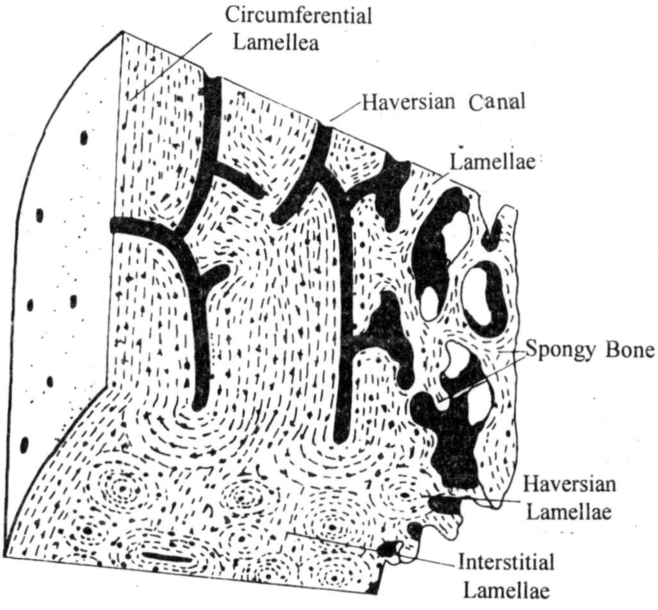

Fig. 20.10 Bone-three dimensional view

Matrix: The matrix of the bone occupies a very large portion. It is very hard and made up of a protein called ossien.. Collagen fibres are also found in the matrix but difficult to observe. The matrix is loaded with salts like calcium phosphate, calcium carbonate, sodium carbonate and magnesium sulphate. Of these the major quantity is of calcium phosphate. The organic and inorganic constituents of the matrix are in the ratio of 30:70. Within the matrix are found a number of layers called lamellae. These are arranged in concentric rings around narrow longitudinal cavities called *Haversian canals*. The Lamellae that surround the haversian canals are called haversian lamellae. Some lamellae however are arranged concentrically around the marrow cavity. These are called circumferential lamellae. Within the lamellae are found small fluid spaces called *bone lacunae*. These give off fine radiating canals called *canaliculi*. Each lacuna has a bone cell or an *osteocyte*. The canaliculi help in communication between different bone cells. Small protoplasmic strands extend from the bone cells and form a network throughout the matrix. The Haversian canal together with lamellae, lacunae and canaliculi forms a Haversian system. The spaces between the Haversian systems are filled with small interstitial lamellae.

There are some transverse canals which connect the Haversian canals. These are called Volkmann's canals. Each Volkmann's canal contains the supply of blood vessels, nerves, a lymph vessel and some bone cells. The blood vessels of the Haversian canal are connected to the blood vessels of periosteum and bone marrow. Nourishment and oxygen to the oestocytes is provided through these blood vessels.

External to the endosteum (lining the inner margin of the matrix) is found a layer of cells called the inner layer of osteoblasts. Similarly the outer margin of the matrix also has a layer of cells called the outer layer of osteoblasts.

The periosteum is a tough sheath made up of white fibrous tissue enclosing the bone externally. It is to this sheath that tendons of the muscles are attached. The periosteum has blood vessels which sends branches into the bone to provide nourishment.

There are some differences between the bone of a frog and that of a mammal. In the case of the frog bone the lamellae are concentrically arranged around the bone marrow cavity where as in the case of mammals the lamellae surround the Haversian cannals.

BLOOD

Blood is a liquid connective tissue. It is referred to as a tissue because it has a number of cells and it is called a connective tissue as the cells are present in a liquid matrix which travels all over the body.

An average adult human being has about 6.8 lts of blood constituting 6 to 10% of the body weight.

Composition of blood: Basically the blood has two portions namely the plasma and the blood cells. The blood cells are of various types.

Blood

- Plasma
- Formed substances
  - Corpuscles
    - Erythrocytes
    - Leucocytes
      - Granulocytes
        1. Basophils
        2. Eosinophils
        3. Neutrophils
      - Agranulocytes
        1. Monocytes
        2. Lymphocytes
  - Platelets

Plasma: The plasma is the fluid portion. It is a light yellow, slightly alkaline liquid. The plasma comprises of 90 to 92% of water and 8 to 9% of solids. The solids include inorganic salts, proteins and a number of other organic compounds.

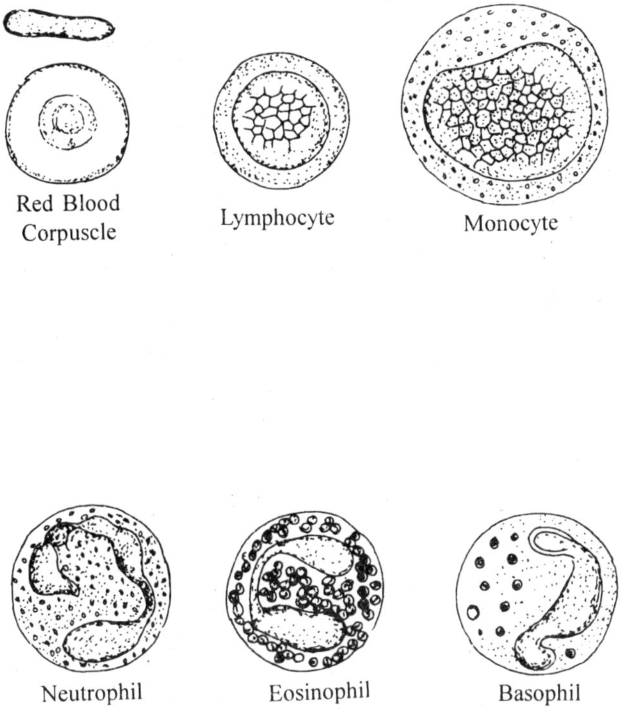

Fig 20.11 Different types of blood corpuscles

Blood cells: These are of basically two types. *Corpuscles* and *platelets*. The corpuscles are again classified into two types- *Erythrocytes* and *Leucocytes* (See classification above).

Erythrocytes: These are commonly called the red blood corpuscles (RBC). As the name indicates they are scarlet red in colour due to the presence of an oxygen carying pigment called Hemoglobin.

The shape of the RBCs varies in different groups of animals. In frog the RBC is biconvex and nucleated; the same is the condition in fishes, reptiles and birds. The RBCs of mammals are circular and biconcave and do not have any nucleus. The central part of the RBCs is thinner and helps in flexibility.

The size of the human RBCs are 7 to 8 mm in diameter. In human beings on an average there are about 4 to 5 million RBCs per cubic mm of blood.

Leucocytes : Leucocytes are commonly called White Blood Corpuscles (WBC). These cells lack haemoglobin and are generally far less in number when compared with RBCs.

Classification of WBCs: Basically there are two types- namely agranulocytes and granulocytes.

(i) Agranulocytes : These do not have any granules in cytoplasm or occasionally small granules may be found. The cells have a single nucleus and are generally rounded in shape. They are produced in Lymphoid tissues. There are two types of agranulocytes viz, *monocytes* and *lymphocytes*.

(a) *Lymphocytes*: These cells are larger than RBCs. They have a large spherical nucleus and thin cytoplasm. They are non motile and also nonphagocytic. They are produced in lymph nodes and secrete antibodies which destroy invading microbes. The lymphocytes are about 1300 to 3000 in number per cubic milimeter of blood.

(b) *Monocytes*: These are the largest of all leucocytes and have a large bin shaped nucleus. They are motile. They come out of the blood vessels and engulf bacteria and other cellular debris.

(ii) Granulocytes : These leucocytes have a number of granules in the cytoplasm and have a distinctly lobed nucleus. The Granulocytes are produced in the red bone marrow. They are also motile and squeeze out of the blood vessels and destroy invading cells. Three kinds of granulocytes have been identified based on their staining reaction. These are basophils, eosinophils and neutrophils.

(a) *Basophils*: These react positively to the basic stains. They are fairly large in size, have a S shaped nucleus and a few coarse granules. A cubic milimeter of blood consists of about 50 cells averaging to

0.4%. The Basophils release the anticoagulant substance.

(b) *Eosinophils*: These leucocytes take up acidic stains like eosin appearing red in colour. They are fairly large, have bilobed nucleus and abundant coarse granules. On an average a cubic milimeter of blood consists of about 60 to 100 cells. Eosinophils destroy toxic substances produced by microbes. In case of infections and allergic reactions the number of eosinophils increases and this condition is called *eosinophilia.*

(c) *Neutrophils*: These exhibit a neutral response to both acidic and basic dyes and stain equally well. These are large, have a many *lobed* nucleus and many *fine* granules in the cytoplasm. They are phagocytic in action.

Platelets:

Platelets are rounded or oval or disklike bodies and lack haemoglobin. Usually a cubic milimeter of blood has about 2,50,000 platelets. Increase or decrease in this number is an abnormality and is known as *thrombocytosis* and *thrombocytopenia* respectively.

Structurally the platelets are flat and bounded by a membrane. The cytoplasm consists of a few organelles and of secretory granules. The platelets are formed in the red bone marrow and survive for about three to seven days only and are disposed off by phagocytosis. The platelets do not have a nucleus. The platelets contain a chemical substance called *thrombokinase* which helps in the clotting of the blood on exposure to air.

Thrombocytes: These are small spindle shaped cells and have a function similar to platelets. These are found in vertebrates other than mammals.

LYMPH: This is a colorless liquid tissue and is made up of two parts viz., *plasma* and *leucocytes.* The plasma is a colorless fluid matrix and consists of floating amoeboid cells having the leucocytes (WBC). The leucocytes are mostly lymphocytes. The plasma does not have RBC, platelets and other blood proteins. The plasma fluid enters from the intercellular spaces into the lymphatic capillaries and through these vessels it finally enters the blood stream.

The lymph transports materials from the tissues into the blood stream. It also destroys the invading microbes.

NERVOUS TISSUE

The nervous tissues are specialised for receiving and responding to stimuli such as sound, heat, cold, chemicals etc. The cells of the nervous tissue help in the transmission of the messages from one to another. The protoplasm of the nerve cells has the property of conductivity through which messages are transferred from one place to another. The nervous system which is composed of nervous tissue has the following three components.

1. Neurons
2. Nerve fibres and
3. Neuroglia

Neurons: Also called nerve cells, the neurons constitute the basic structural unit of the nervous tissue. Each neuron is made up of three parts namely the cyton, dendrites and axon.

(i) *Cyton* : This is also called the cell body, soma or perikaryon. Body of the neuron may be rounded, oval, or stellate. The cytoplasm is abundant and it is called neuroplasm. The cyton has a large nucleus with a distinct nucleolous. The cytoplasm also has mitochondria golgi aparatus, flat globules and fine thread like structures called neurofribrils. There is no centrosome so that the neuron cannot divide. But some zoologists opine that centrosomes are present in the neurons. In addition to this the cytoplasm has numerous basophilic bodies called *Nissle's granules*. The exact functioning of these granules is not understood. But they seem to play an important role in the metabolism of the neurons as indicated by the fact that they undergo changes

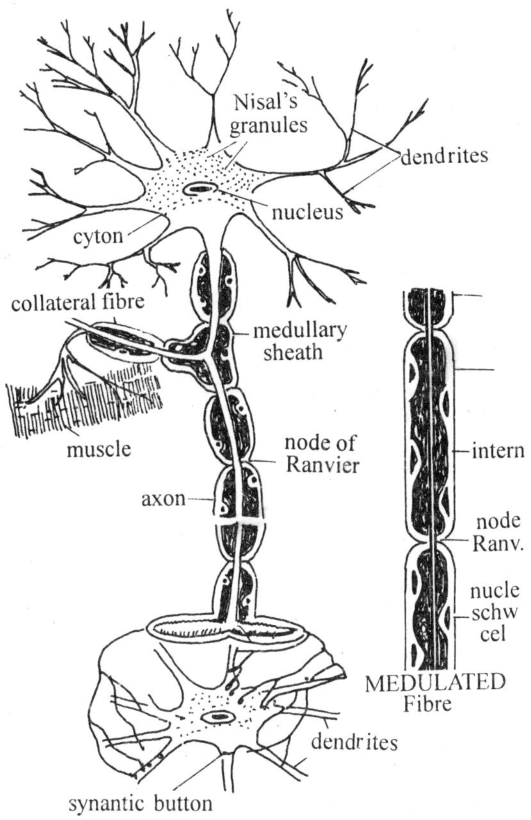

Fig 20.12 Structure of a neuron

under different physiological conditions. Electron microscopic studies of the Nissle's granules has shown it to be rough endoplasmic reticulum.

(ii) *Dendrites* : These are small minute processes that arise from tne cyton. They show a high degree of branching. The function of these dei drites is to connect one neuron with the another.

(iii) *Axon* : This is a single long process that arises from the cyton. It is usually given off from the small conical region of the cyton which is known as axon hillock or implantation cone. The cone contains ribonucleic acid, protein and iron. The axon at its free end gives rise to minute branches called telodendrites. The cytoplasm of the axon is known as axoplasm.

CHAPTER 21

ANIMAL REPRODUCTION

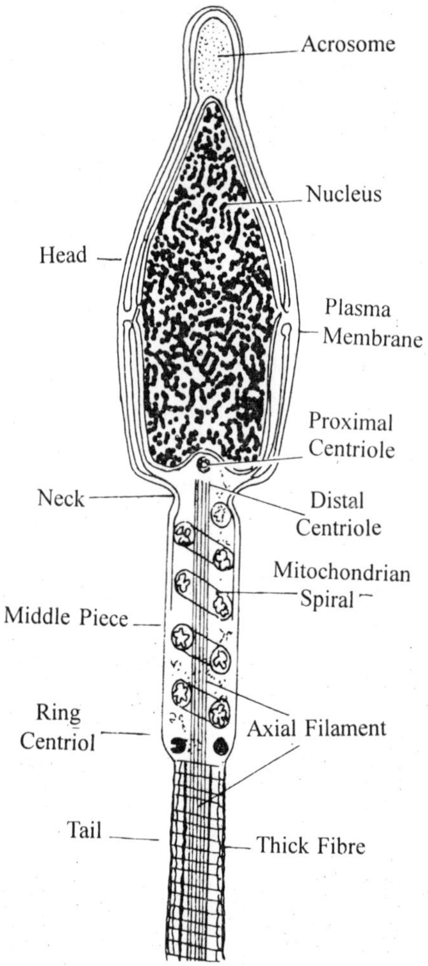

Fig 21.1 A fully mature sperm

STUDY OF GAMETES

Structure of the sperm

The sperm in human beings is a highly specialised structure. A mature sperm consits of the following four parts -

1. **Head.** It is conical in shape and is formed of acrosome and nucleus. The acrosome is found at the anterior tip of the sperm and forms the head cap. The head cap occupies the space between antierior half of the nucleus and the plasma membrane of the sperm tip. The acrosome is bound by the membrane. The sperm nucleus occupies most of the available space of the sperm head. Structurally the nucleus is enveloped by a membrane. Some times the posterior part of the nuclear membrane is somewhat depressed to accomodate the proximal centriole.

2. **Neck.** The head is followed by a short neck to separate the middle piece of the sperm. The neck consists of two granules (centrioles). These granules are called the proximal centriole and the distal centriole.

3. **Middle piece.** This conists of the upper portion of the axial filament and is made up of two longitudinal fibres called beta fibres and alpha fibres. The alpha fibres are nine in number. In human sperms the alpha fibres are further accompanied by nine thicker fibres called gamma

fibres. These fibres are surrounded by the mitochondria. Around the periphery of the middle piece of sperm a thin layer of cytoplasm is present.

4. Tail. This is the longest part of the sperm. In human beings it is about 55microns longs. It conists of two main parts the principle piece and the end piece. The principle piece which constitutes most of the length of the tail is made up of axial filaments with a 9+2 arrangement. The end piece is a small tapering portion of the tail containing only the axial filament covered with cytoplasm and plasma membrane.

Structure of a mature human egg (Graffian follicle)

A mature human egg is called Graffian follicle. The young ovum develops and undergoes maturity in the follicles of the ovary called Graffian follicles. The mature egg has a typical cell structure and consists of the following parts -

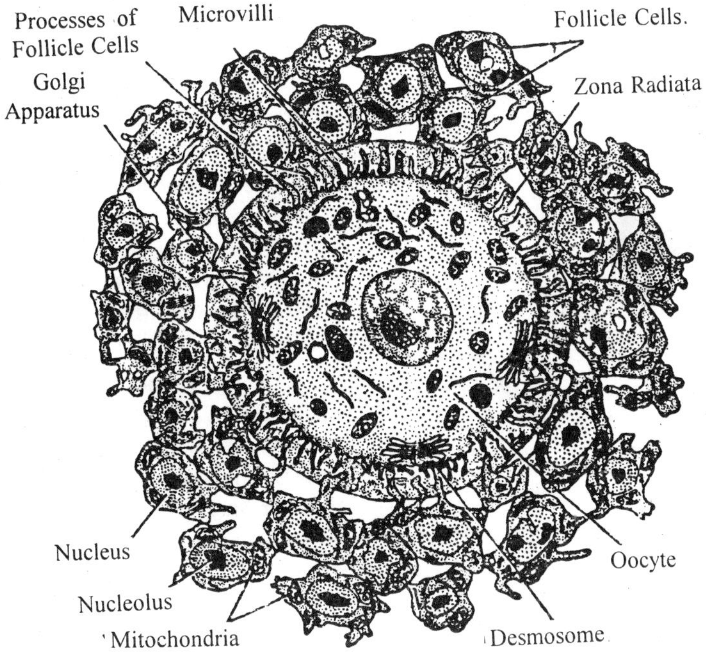

Fig. 21.2 Young oocyte of a mammal surrounded by follicle cells

The outermost layer is a double layered connective tissue sheet and it is differentiated into two layers namely *theca externa* and *theca interna*. The theca externa is the outer fibrous layer. It is made up of spindle shaped cells and fibres. The theca interna is composed of the inner vascular layer and it is supplied by a number of blood capillaries. In a mature follicle there is a fluid filled cavity which is eccentric and it is called antrum. The fluid itself is called liquor follicle. The antrum is lined by a membrane called membrana granulosa. Between the theca interna and membrana granulosa there is a homogenous basement membrane called glassy membrane.

Internal to the theca interna is a membrane called zona pellucida. This is comparable to the vitelline membrane. Surrounding the zona pellucida is a region called corona radiata. This is formed from the other follicle cells.

The cytoplasm of the egg is known as the ooplasm. The ooplasm consists of a large amount of reserve food materials. There is very little yolk however in the cytoplasm. The peripheral part of the ooplasm is known as the cortex. It consists of many microvilli. The microvilli are formed as protruberances of the plasmalemma (plasma membrane). The microvilli help in the transportation of the substances from follicle cells to the egg during its development. The cytoplasm consists of a number of mitochondria, lipoproteins, pigment granules, ribosomes, golgi complex and a number of other cell inclusions. The egg consists of a single nucleus which is usually vesicular in shape.

The egg of frog

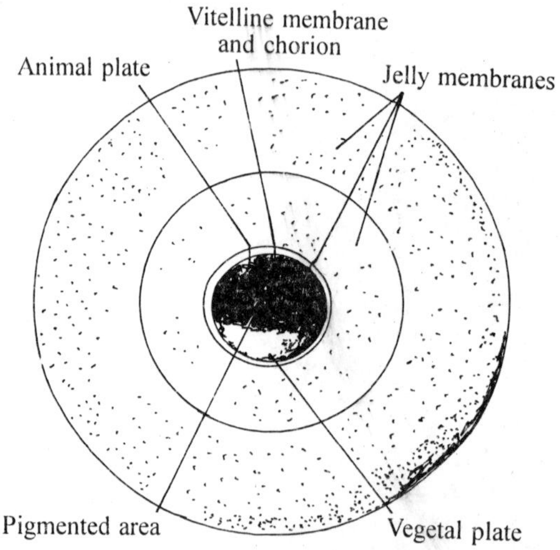

Fig. 21.3 Frog's egg

The egg of frog is of the mesolecithal type. Here the yolk is located in the vegetal pole of the egg. While the animal pole conists of cytoplasm and nucleus. The ooplasm of egg consists of many pigmented granules present in the upper region of the egg. The vegetal pole does not have any granules and hence creamy in colour.

The outermost envelope of the egg is a jelly covering and it made up of three layers. These are the outer tough and sticky layer, the middle fluid like layer and an inner thin layer. Next to this is the vitelline membrane which is non cellular.

The egg proper is minute in size and has a diameter of 2mm. It is spherical in shape. There are two regions in the egg the vegetal pole and the animal pole. The vegetal pole is creamy in colour whereas the animal pole is blackish brown in colour because of the presence of granules.

The egg contains a moderate amount of the yolk hence it is said to be mesolecithal. There is a less pigmented region between the vegetal and animal poles. This is called the grey cresent. The grey cresent region develops after the entry of sperm.

EMBRYOLOGY

Study of developmental stages of frog

The egg of frog before fertilization will have a radial symmetry. After fertilization however, this changes into bilateral symmetry due to the penetration by the sperm. The sperm as it enters the egg carries along with it certain granules. As a result of this many pigment granules are carried away from the animal pole. Due to this, a less pigmented area is created at the region opposite to the point of entry of the sperm and this is called the grey crescent. The presence of the grey crescent forms the posterior side and the opposite side forms the anterior side of the future embryo. At this stage there will be two planes in the egg. One is the sperm enterance point plane and the other is grey crescent plane.

At the initiation of the first division (cleavage) another plane is produced known as the cleavage plane which usually passes through both the animal and the vegetal poles.

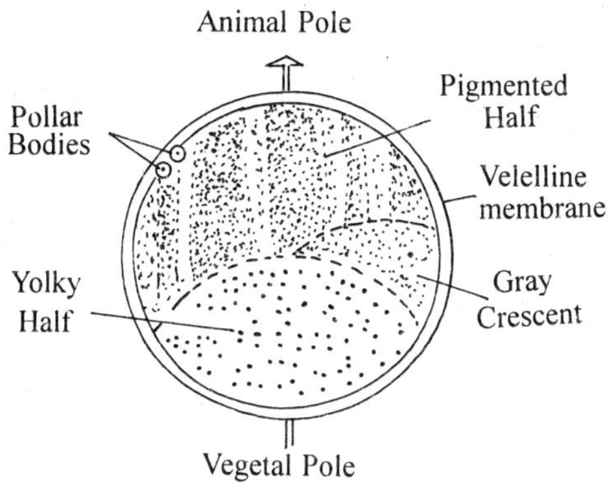

Fig. 21.4 T.S. of Zygote

The early development of the zygote consists of three main stages - cleavage, blastula and gastrula. At the end of this stage the primary germ layers are formed and from this develop all the further organs of the frog.

Cleavage

The first division of the zygote is said to be the cleavage or segmentation. This is a mitotic division. The cleavage is said to be of the holoblastic type. The first cleavage results in the formation of two unequal cells. Hence the cleavage is said to be of the unequal type. The pigmented half contains cells that are smaller in size than the half containing the yolk.

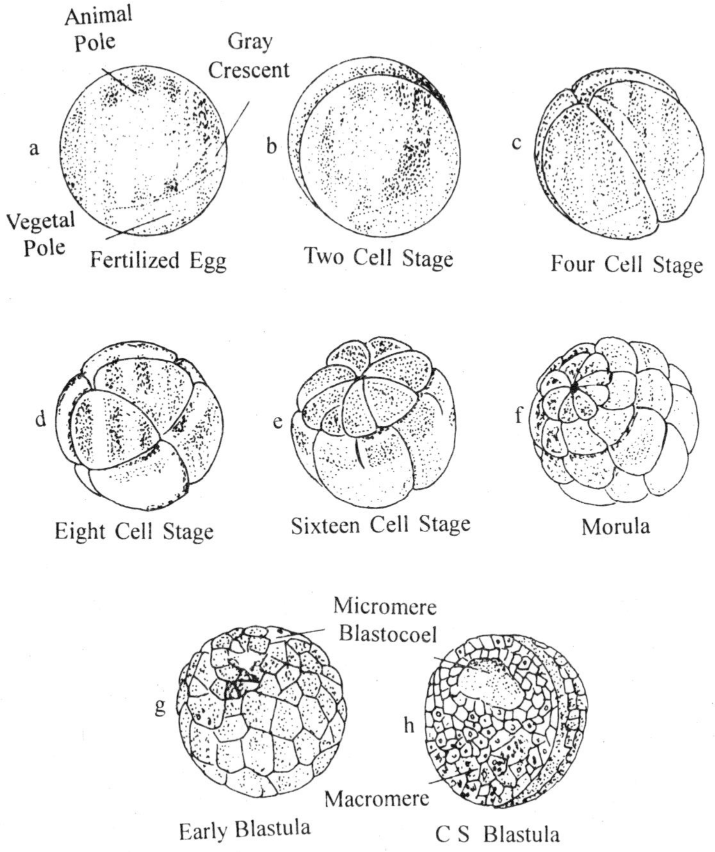

Fig. 21.5 Cleavage in fertilized egg or zygote

The first cleavage in a zygote of frog is meridional ie., it passes through both the animal and vegetal poles. The two cells formed as a result of the first cleavage are called blastomeres.

The second cleavage takes place after about one hour of the first cleavage. This cleavage is also meridional but it takes place at right angles to the first plane of division. As a result of this division four blastomeres are formed. Of these, two contain the grey cresent and the other two are without it.

The third cleavage which begins 80 minutes after the second cleavage is horizontal and is at right angles to the second cleavage and passes slightly about the equator. The third cleavage results in the formation of eight cells. Of these eight cells four cells towards the vegetal pole are larger in size and have yolk content. These are called megameres. The remaining four cells towards the animal pole are smaller in size, pigmented and are called micromeres.

The fourth cleavage is meridional and it consists of two cleavage planes passing between the first and the second cleavage. As a result of this cleavage 16 cells are formed, of which 8 are megameres and the other 8 micromeres.

The fifth cleavage is double and horizontal. It consits of two latitudinal cleavage furrows. One of these cleavages is above and the other below the third cleavage furrow. As a result of this 32 cells are formed of which 16 are pigmented micromeres and the other 16 yolky megameres.

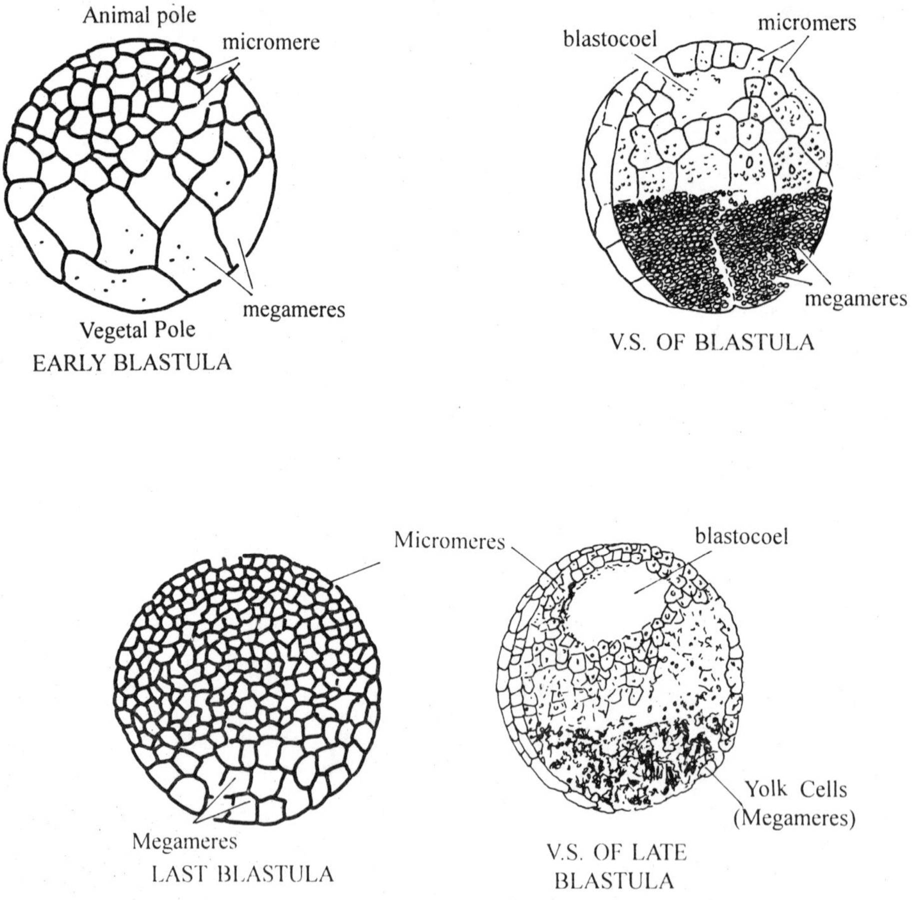

Fig. 21.6 Early and late blastula

From this stage onwards the division is rather irregular and forms a ball like structure called the morula.

Blastula

At the end of cleavage the solid ball of cells gives rise to blastula which consists of a number of blastomeres. The characteristic feature of the blastula stage is the presence of a well defined cavity the blastocoe. This is the beginning of the primary body cavity. The process of the formation of blastula is called blastulation. The blastula of frog is called amphiblastian as the cavitiy is confined to only the animal pole. The vegetal pole however is composed of a solid mass of non pigmented yolky cells.

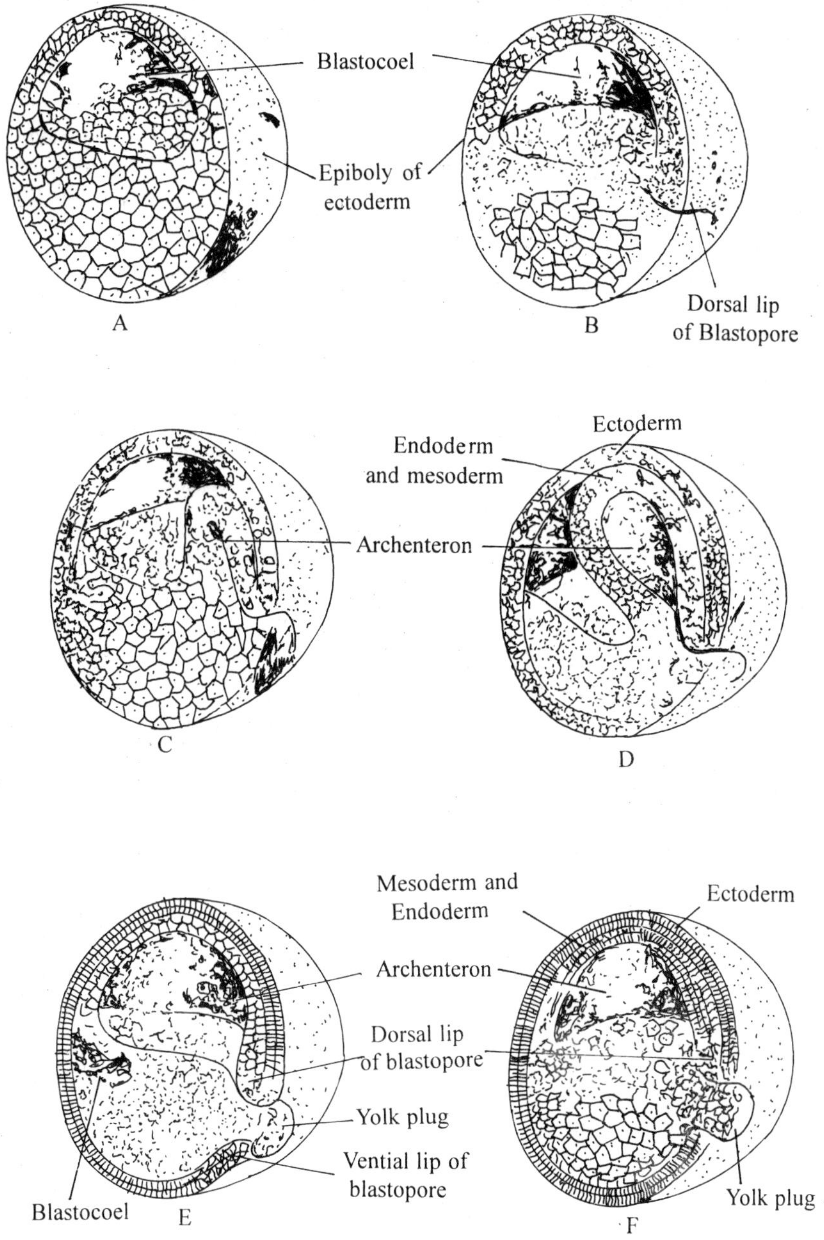

Fig. 21.7 The blastula (A) and its transformation into gastrula (B-F)

In the 32 cell stage, the blastula consists of a single layer of cells but as the segmentation proceeds the number of cells in the blastula increases. The blastocoel also enlarges in size. Between the micromeres and the megameres are found a group of intermediate cells constituting the germ ring. The pigmented micromeres develop into the ectoderm while the germ ring gives rise to the mesoderm. The megameres give rise to the endoderm. The outer most layer of cells in the blastula is ciliated. This helps in the rotational movement of the blastula.

Gastrula

The process of formation of gastrula is called gastrulation. In the beginning the gastrula is a two layered stage with a cavity called archenteron. This archenteron opens to the exterior by means of an opening called the blastopore.

As a result of gastrulation three primary layers are formed. These are the ectoderm, endoderm and the mesoderm. Gastrulation is a very significant and important stage in the embryogeny of the frog because it is here that all the future organ forming areas are differentiated. Gastrulation consists of the following stages -

1. *Epiboly.* During epiboly the ectoderm overgrows backwards on the endoderm ultimately the entire embryo (except for the small area called the yolk plug) is covered by the ectoderm.
2. *Invagination.* A small depression is formed in the region occupied by the grey crescent area. This depression grows inwards and forms the gastrocoel or the secondary body cavity.
3. *Involution.* During this process the cells which have grown backward during epiboly now roll inside at margin of the blastopore.
4. *Closure of the blastopore.* The archenteron at its beginning is marked by a groove. This groove extends laterally and is completely ring like. The margins of the lip of the blastopore contract reducing the size of the opening.
5. *Rotation.* At the end of gastrulation the yolk filled cells get rearranged. This rearrangement shifts the centre of gravity of the gastrula and results in the rotation of 80-90°. The blastopore moves backwards and is finally located slightly above its original location.
6. *Convergence.* During this process there are two kinds of cell movements. First there is shifting of the cells of the notochord and mesoderm towards the dorsal lips. Secondly there is a dorsal convergence of notochord and mesodemal cells towards the blastopore.
7. *Extension or elongation.* This includes the growth in length of the embryo. As a result of extension the spherical gastrula becomes elongated in shape.

CHAPTER 22

HUMAN ANATOMY

Structure of the human heart

The heart is covered by a wall called pericardium which is made up of three layers. These are the epicardium, the myocardium and the endocaridum. The epicardium is a serous membrane and is the outer most layer. This secretes a serous fluid into the space between the visceral and the parietal pericardium. This fluid allows smooth movement when the heart beats. It has been estimated that there will be normally about 15 ml of the pericardial fluid. The myocardium is the intermediate layer. It is composed of specialised muscle tissues called the cardiac muscles. These muscles are striated fibres. The myocardium is thin in the auricles but thick in the ventricles specially so in the left ventricle.

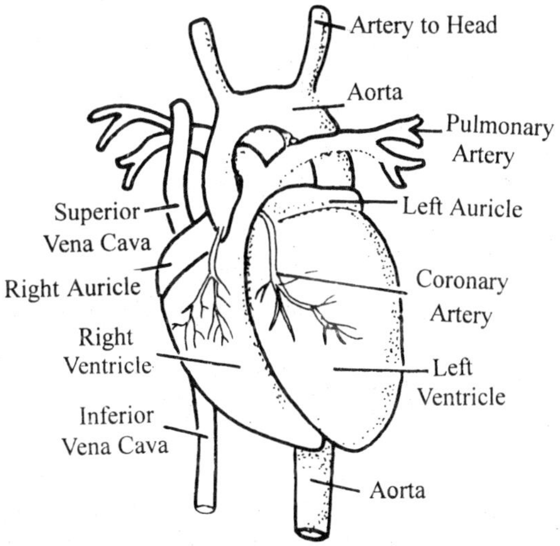

Fig. 22.1 Human heart

The endocardium is the inner most layer. This is a thin membranous structure in contact with the blood. It is composed of flattened endohelial cells. The endocardium is thicker in the auricles than in the ventricles.

Internally the heart is composed of 4 chambers namely

1. Right auricle (atrium)
2. Left auricle (atrium)
3. Right ventricle and
4. Left ventricle

There are two auricles - the right and the left. These are separated by a thin interauricular septum. The ventricles are also two in numbers and they are separated by the interventricular septum. The right auricles open into the right ventricle by the right auriclo ventricular openings guarded by a tricuspid valve having three flaps. The left auricle opens into the left ventricle by left auricul ventricular aperture guarded by a bicuspid valve having two flaps. This valve is also called the mitral valve.

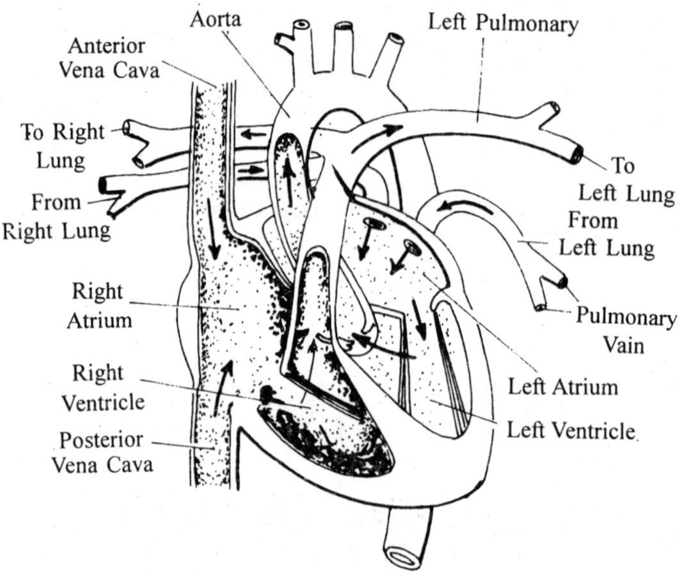

Fig. 22.2 Vertical section of Heart

The right auricle receives venous blood through superior and inferior vena cave. The left auricle receives oxygenated blood through the pulmonary vein returned from the lungs.

STRUCTURE OF THE BRAIN

Brain is the controlling centre of all the body activities. It is protected by cranium. It is covered by three membranes together known as meninges. The inner membrane is piamater the middle arachnoid and the outer membrane is duramater. Outer part of the brain has grey matter and its inner part has white matter. Inside the brain there are 2 spaces - subdural space which is the space between duramater and arachnoid and subarachnoid space which is the space between arachnoid and piamater. These are continuous with the spinal canal of the spinal cord and are filled with cerebro-spinal fluid. The human brain weighs about 1,400 gms in volume.

Divisions of human brain

Human brain has three primary divisions (with subdivinions)

Forebrain, Midbrain and Hindbrain.

Forebrain. It is the largest part of the brain constituting about 2/3 of the human brain. The forebrain is distinguished into cerebrum, olfactory lobes and diencephalon.

1. *Cerebrum*. The cerebrum is the largest and most prominent part of the brain. It is divided into the right and left cerebral hemispheres. Each hemisphere is divided by sulci into distinct areas known as frontal, parietal, temporal and occipital lobes. Ventrally, the two cerebral hemispheres are connected by a transverse band of white medullated fibres the corpus callosum. Its outer wall is the cerebral cortex which is thrown into folds forming elevations called gyri with depression between them termed as sulci.

2. *Olfactory lobes*. These form the anterior most part of the forebrain and are more distinct on the ventral side.

3. *Diencephalon*. It is a small structure covered dorsally by cerebral hemispheres. Its dorsal surface is raised into a pineal stalk bearing a small rounded pineal body. From the floor of diencephalon arises a funnel-shaped pituitary body. The dorsal wall of diencephalon is thin and vascular and represents the anerior choroid plexus. The cavity of diencephalon constitutes the third ventricle. It communicates through the Foramen of Monro with the cavity of cerebral hemispheres.

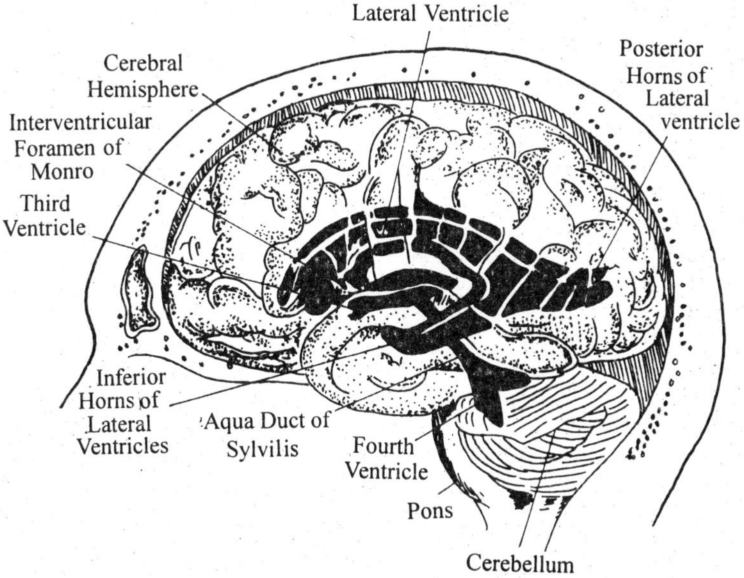

Fig. 22.3 Sagittal section of human brain

Midbrain : The midbrain consists of optic lobes and cura cerebri.

1. *Optic lobes*. The optic lobes are four solid lobes lying behing the diencephalon. They collectively constitute the corpora quadrigemina. The anterior two lobes are slightly larger than the posterior lobes. The anterior and posterior lobes are connected by means of posterior commissure. They are devoid of optic ventricles, but between them passes a narrow cavity, the iter, which connects the third and fourth ventricles together.

2. *Cura cerebri*. These are two longitudinal bands of nervous matter situated on the ventral side of the optic lobes which connect the medulla oblongata with the cerebral hemisphere.

1. *Cerebellum*. The cerebellum is large and projects anteriorly upto the cerebrum. Its surface is produced into numerous folds. The cerebellum consists of a central lobe, the vermis, two lateral lobes and two floccular lobes attached to the lateral lobes. The vermis is divided by transverse furrows into anterior, middle and posterior portions. Its grey matter presents a branched tree like structure which is known as arbor vitae.

2. *Medulla oblongata*. It is the posterior most part of the brain and is located underneath the cerebellum. Its floor is beset with a transverse broad band of fibres of white matter which is known as pons varolii. The anterior triangular part of medulla possesses thick floor and lateral walls but its roof is exceptionally thin and is formed of highly vascular piamater. It represents the posterior choroid plexus. It is perforated by a small aperture which is known as foramen of Magendie. Its cavity is known as metacoel.

The **hind brain** consists of cerebellum and medulla oblangata.

HUMAN ANATOMY – STUDY OF RESPIRATORY SYSTEM

The respiratory system in human beings consists of lungs and air passages. In its course to the lungs the air passes through the nose, mouth, pharynx, larynx, trachea, bronchi and ultimately the lungs.

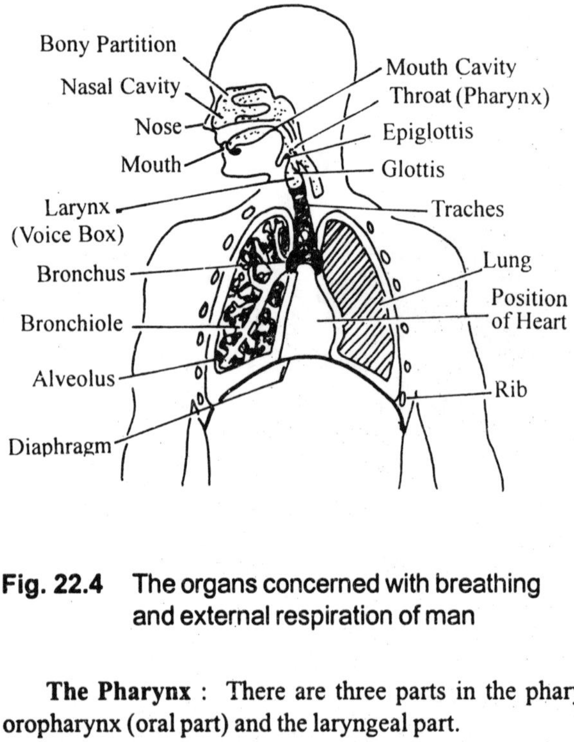

Fig. 22.4 The organs concerned with breathing and external respiration of man

The nose has the external(anterior) nares, nasal chambers and the internal (posterior) nares.

External nares : The paired external nares are situated at the anterior end of the upper jaw slightly behind the mouth. These are rounded in form and communicate posteriorly with paired nasal chambers.

Nasal chambers : The nasal passages are paired elongated chambers situated dorsal to the buccal cavity and separated from the latter by bony palate. The nasal passage opens anteriorly by external nares to the outside and posteriorly it leads into the pharynx through the nostril.

The internal nares : These are situated at the back of the nasal cavity and constitute the enterance to the nasopharynx.

The Pharynx : There are three parts in the pharynx. These are the nasopharynx (nasal part), the oropharynx (oral part) and the laryngeal part.

The Larynx : The larynx is situated in the anterior part of trachea and communicates with the pharynx through glottis. Besides acting as a part of the air passages, the larynx is modified in structure to help in voice production.

The larynx consists of a framework of the Thyroid cartilage, Cricoid cargilage and Arytenoid cartilages.

Epiglottis: This is a leaf shaped plate made up of yellow elastic fibrocartilage and is situated between the base of the tongue and the upper opening of the larynx. The lower portion of the epiglottis is attached through a ligament to the thyroid cartilage. The main function of the epiglottis is to prevent the entry of food to the larynx during swallowing.

Hyoid: This is a horseshoe shaped bone lying between the mandible above and larynx below. It is located at the base of the tongue and provides attachment to this. This however does not take any part in the formation of the larynx. Internally the larynx is lined by the mucous membrane which is covered with ciliated columnar epithelium except over the vocal cords.

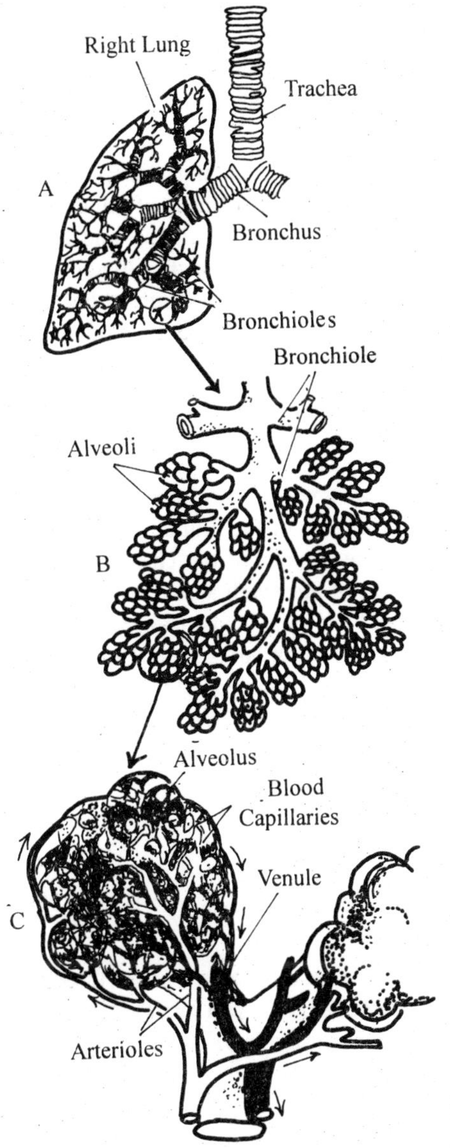

Fig 22.5 (a) Lungs in man (b and c) Structure of bronchiole and alveolus

The Trachea: Commonly called the windpipe the trachea is about 12 cms long and about 2.5 cms in diameter. The upper half of the trachea is situated in the midline of the neck while the lower half is in the superior mediastinum of the thorax. It lies in front of the oesophagus and ends opposite the fourth thoracic vertebra. Here it divides into two branches called bronchi.

The Bronchi: The trachea bifurcates into two main branches called the bronchi (right and left) opposite the fourth thorasic vertebra. Each bronchus enters to respective lung (left bronchus to left lung; right bronchus to right lung). Each main bronchus divides into numerous smaller branches repeatedly and the ultimate small tubes of the bronchus are called *bronchioles*.

In structure, the bronchi are similar to trachea; they are surrounded by incomplete hyaline cartilages lined with mucous membrane made up of ciliated columnar epithelium. The walls of the bronchi also have some smooth involuntary muscles. The bronchioles have a structure similar to bronchi but there are no cartilaginous structures; instead they consist of muscle fibres.

The Alveoli or Air sacs: The Bronchioles ultimately terminate in irregular sac like structures made up of a number of air packets called *alveoli* or *air sacs*. The alveoli are lined with a delicate layer of epithelial cells and are surrounded by numerous capillaries. It is through the walls of these capillaries that the exchange of air takes place.

The wall of the air sacs is made up of three layers, namely a surface epithelium, a supporting connective tissue layer and blood vessels.

The Lungs : In human beings these are the main respiratory organs. These are a pair of conical organs enveloped by a serous membrane - the *pleura*. The lungs occupy a greater part of the thorasic cavity. The two lungs are separated from each other by the mediastinum. Each lung is divided by deep fissures into lobes. The right lung has three lobes - the upper, middle and lower while the left lung has two lobes the upper and the lower. The lung is also described as having a mediastinal and a costal surface, an apex and a base.

The outer(costal) surface is in contact with the wall of the pleural cavity while the medial surface of the lung is applied to the mediastinum. The main feature here is the presence of hilum where the main bronchus and pulmonary artery enter and the pulmonary veins leave the lung. Lymph nodes are present at the hilum which may get enlarged by diseases.

The apex of the lungs rises into the root of the neck for about an inch above the clavicle. The base is concave and is related to the upper part of the diaphragm.

Structurally the lungs are spongy and each lobe of the lung is made up of a number of small lobules consisting of alveoli and bronchioles which form larger bronchi. In addition to this there is a framework of interstitial or fibrous tissue in which are found the blood vessels and lymphatics.

The pleura: This is a serous membrane and is made up of two layers - the *parietal* and the *visceral*. The visceral layer forms the outer cover of the lung which it encloses completely except at the region of the hilum. The parietal layer lines the interior of the chest wall and upper surface of the diaphragm. Both the layers are smooth and shiny and are always kept moist by a small amount of serous fluid acting as a lubricant. This helps the two surfaces to glide smoothly over each other during the process of breathing.

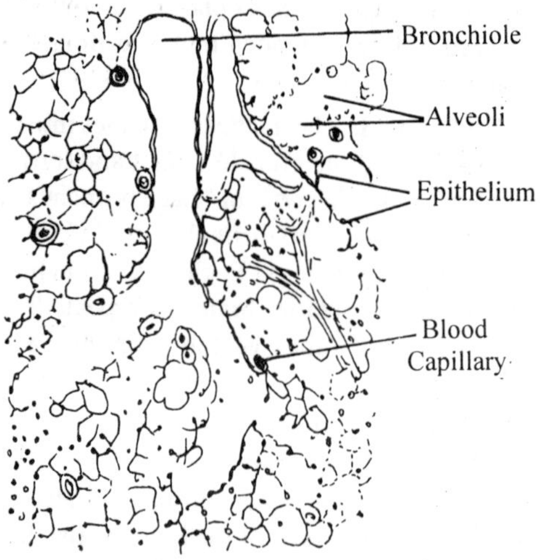

Fig 22.6 T.S. of Human lung

EXCRETORY SYSTEM OF MAN (MAMMAL)

Kidneys : These are the most essential organs of man and other mammals. These are 2 in number, and are dark red in colour placed one on either side of the vertebral column.

Each kidney is convex on the outer surface and concave on the inner side termed hilum through which renal artery enters and renal vein and ureter leave. The ureters lead behind to open into the urinary bladder and are continued as urethra which opens at the tip of the penis in the man and in front of the vaginal opening in the female.

STRUCTURE OF THE KIDNEY

In human beings the kidneys are paired. They are bean shaped and are located between the levels of last thoracic and third lumbar vertebrae. The right kidney is placed slightly lower than the left because of the presence of the liver.

The average size of the human kidney measures about 11 to 13 cms in length; 5 to 7.5 cms wide and 2.5 cms thick. In adults each weighs about 150 gms and this correspondence to about 1% of the body weight. The medial border of the kidney is slightly concave and has a marked depression.

A long section of the kidney shows two main regions - an outer reddish cortex and an inner pale coloured medulla.

Cortex: The Cortex is granular in appearance because of the presence of convoluted uriniferous tubules. The cortex overlies the bases of the medullary apices called medullary pyramids. The cortex is also present between the medullary pyramids. Under a magnifying glass the cortex can be seen divided into alternating radial tracks known as *rays* and *labyrinths*. The rays are radially straight because they contain the straight parts of the uriniferous tubules. The Labyrinths have a granular appearance because the tubules are cut irregularly.

Medulla: The medulla looks striated as the tubules run through it in a straight course. It consists of 10 to 15 conical masses called the renal pyramids. Two or three pyramids usually fuse and form the common apex called papilla. These papillae project into cup shaped channels known as calyces. The calyces join to form the pelvis which is the expanded initial part of the ureter.

NEPHRON (URINIFEROUS TUBULE) This is the basic filtering unit of the kidney. It has been estimated that nearly one million nephrons are present in the human kidney. The nephron basically consists of two parts. The secretory tubule and the excretory portion.

1. Secretory tubule. The secretory tubule consists of the Bowman's capsule, the proximal convoluted tubule, the loop of Henle and distal convoluted tubule.

Excretory tubule. This is divided into two regions namely a collecting tubule and a papillary duct.

Collecting tubule : The distal convoluted tubule straightens and by a short junction tubule joins the collecting tubule. The collecting tubules have a lining of cuboidal epithelium. These tubules pass down a medullary ray and enter into the medullary pyramids. On its way, the tubule receives other junctional tubules and forms a much branched tree like system.

Papillary duct : The collecting ducts finally fuse to form a large straight tubule - the papillary duct which opens into the pelvis at the apex of the papilla.

HISTOLOGY–STRUCTURE OF LIVER AND PANCREAS

Liver. The liver is the largest gland in the body weighing about 1.6 kgs. It is reddish brown in colour and it is covered with a connective tissue capsule. The liver is heavier in the male than in the female. It is a wedge shaped organ lying immediately below the diaphragm in the right hypochondrium and epigastrium. The liver has two lobes namely the large right lobe and the small left lobe. Besides this it has superior, inferior, anterior and posterior surfaces.

The right lobe is much larger than the left anu division between them is marked by falciform ligament on the anterior surface. The under surface of the right lobe is further subdivided into two smaller lobes called the *quadrate lobe* and *caudate lobe*. A thin walled pyriform sack like structure called the gall bladder is present in a groove in the inferior surface.

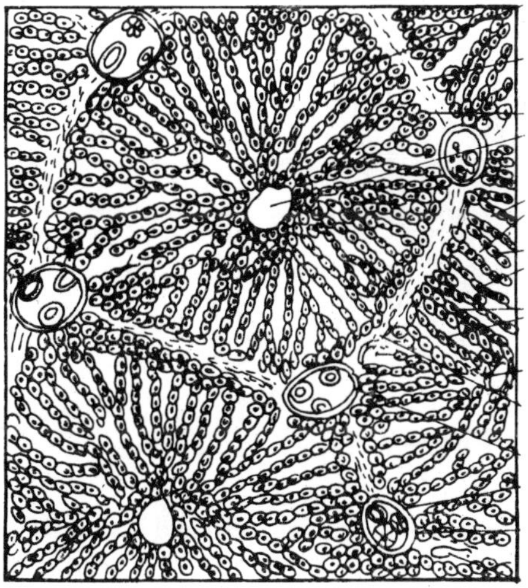

Fig. 22.7 Section of liver

Structure of the liver. The liver is composed of a large number of hexagonal lobules. Each lobule is about 1mm in diameter. At the centre of each lobule runs a small branch of the hepatic vein. The cells of the liver are arranged in one celled thick sheets around the central vein. There are certain spaces between the cells of these plates wherein lie sinusoids. These sinusoids are blood vessels with incomplete walls. They are irregular in shape and wider than blood capillaries. They are lined by certain specialised cells called *Kupffer cells* which are phagocytic in action and remove cell debris and bacteria in the blood. Each lobule receives the branches of hepatic artery and the portal vein. The blood vessels form small branches and enter the sinusoids.

The liver cells are polyhedral in shape and have two surfaces that face the sinusoids and several surfaces in contact with other cells. The most important secretion of the liver is the bile. This is secreted by the liver cells and drains into minute cannals called *bile canaliculi* which lie between the walls of the cells. The bile canaliculi join to form intralobular ductules which unite to form the right and the left hepatic ducts. These two ducts unite to form the common hepatic duct. This runs downwards for about three centimeters where it is joined by the cystic duct which comes from the gall bladder and continues downwards as the common bile duct. This common bile duct enters the duodenum at a small papilla called the ampulla of Vater.

The bile consists of bile pigments and bile salts. The salts present in the bile are sodium glycocholate and sodium taurocholate. The bile pigments are bilirubin and biliverdin. Bile is a golden yellow bitter fluid secreted by the human body to the tune of 500 to 1000 ml. daily; about 86% of it is water.

Pancreas

The pancreas is both an exocrine and an endocrine gland. As an exocrine function it secrets many digesting enzymes such as pancreatic lipase. As an endocrine function it secrets two very important hormones *insulin* and *glucagon*.

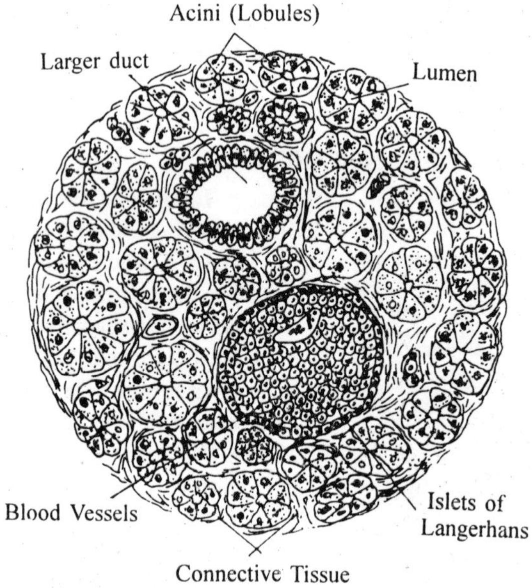

Fig. 22.8 Section of pancreas

The pancreas is a grayish pink gland that lies across the posterior abdominal wall and is situated behind the stomach. The pancreas is supposed, to have three parts. A head, a body and tail. The head is situated at the right and fits into the curvature of the duodenum. The body lies behind the stomach and tail of the gland is in contact with the gastric surface of the spleen.

Structurally the pancreas consists of a number of lobules supported by fine connective tissue. Each lobule contains groups of secretory cells called acini. From the lobules arise small ducts which unite to form large ducts untill they reach the main pancreatic duct. This duct extends from left to right and connects itself to the common bile duct to enter the duodenum at the ampulla of vater. Embedded between the acinar cells lie clusters of special cells called the *islets of Langerhans*. There are two types of cells in the islets of langerhans namely the *alpha cells* and the *beta cells*. The alpha cells secret an enzyme called *glucagon* which is concerned with the break down of the glycogen in the liver. It increases the glucose level in the blood. The beta cells secret *insulin* which controls the sugar metabolism and maintains the normal sugar level in the blood.

CHAPTER 23

LIFE CYCLE OF SILK WORM

STUDY OF LIFE HISTORY OF SILK WORM MOTH

The most popular among the silk moths is *Bombyx mori* which is popularly called the mulberry silkworm, as its main feed is mulberry plant. The life history of mulberry silk worm consists of four stages- the adult, the egg, the larva(catripillar) and the pupa stage.

Adult stage - The adult silk moth is about 2.5 cms in length and pale creamy or white in colour. The moths are dioecious (male and female moths are separate). In the female, due to the heavy body and generally weak wings flight is impossible. The female moths are larger than the male.

The body of the moth is divided into three parts namely head, thorax and abdomen. The head has got a pair of compound eyes and there are two antennae. There are three segments in the thorax- these are the prothorax, the mesothorax and the metathorax. The prothorax is quite big. The moth has got three pairs of legs and there are two pairs of wings.

The female moth generally does not feed during its very short life of about 2-3 days.

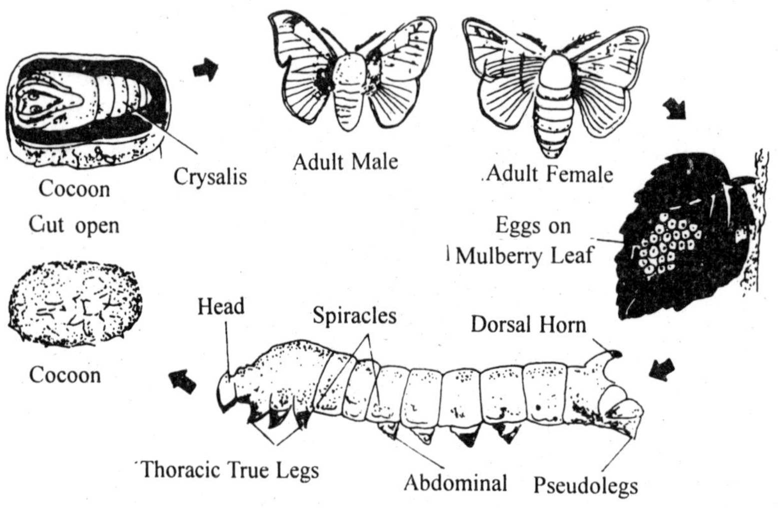

Fig. 23.1 Life history of silk moth

Fertilization- After copulation, fertilization takes place. Just after the emergence from the pupa, the male moth copulates with the female for about 2-3 hours and then they separate. After fertilization the female starts laying eggs and it takes about 24 hours to complete the process. On an average a single moth can lay about 400-500 eggs. The quantity of eggs depends on the climatic conditions and the supply of food material to the caterpillar from which the female moth is developed. They (eggs) are usually laid on the upper surface of the mulberry leaves. The eggs are usually in clusters and are enclosed in a gelatinous substance which is secreted by the female moth. The gelatinous substance helps in the attachment of the eggs to the leaves.

The eggs are minute in size, spherical in shape and white in colour. The weight of about 1000eggs will be less than one gram. Gradually the eggs become dark in colour. It has been observed that in mulberry silk worms there are two types of eggs. These are the *diapause type* and the *nondiapause* type. The silk worms of temperate regions lay the diapause type whereas the silk worms of subtropical regions (as in India) lay the nondiapause type of eggs. In diapause type of eggs there is a hibernating stage (resting stage). The nodiapause type of eggs are nonhibernating and consequently the moth will take a very short time to complete the life cycle. In fact, the mulberry silk moth in India can complete five life cycles per year.

Hatching of eggs - After 10 days of incubation, the eggs are hatched into larvae called the caterpillar. The hatching stage is the most critical stage in the life history of the silk moth. The caterpillars are voracious eaters and need a continuous supply of food (mulberry leaves). If there is insufficient supply of food the caterpillars will not grow properly. Sometimes due to lack of food the larva may even die.

Caterpillar (larva) - The newly hatched larva is about 1/8th of an inch in length. The body is yellowish white in colour. The larvae are provided with well developed mouth parts which are adopted to feed on the leaves of mulberry . In structure, the larval body is made up of 12 segments and the abdomen is made up of 10 segments and has 5 pairs of pseudo-legs. In addition to this, in the last anal segment there is a small dorsal horn.

The silk worms undergo moulting(ecdysis) four times and the stages inbetween the moultings are called *instars*. There are 5 *instars* in the larval stage of silk moth.

The first instar larva can feed only on tender leaves of mulberry as it is at a very young stage. The first instar larva feeds continuously and grows rapidly. After the first, second, third and fourth moultings the larva grows into second, third, fourth and fifth instars respectively. The whole moulting process upto the 5th instar stage takes about 21-25 days after hatching.

The 5th stage larva or the caterpillar can be regarded as the final stage of growth. The caterpillars at this stage are fully grown and measure about 7.5 cms in length and in weight it ranges between 4-6 gms. The caterpillars develop salivary glands and generally stop feeding before they reach the pupa stage.

Pupa - This is the next stage of development. Initially the fully grown caterpillars stop eating and gradually withdraw towards a corner among the leaves in the feeding trays. They secrete a sticky fluid through the silk glands. The fluid comes out through a narrow pore located on the hypopharynx of the caterpillar. The fluid gradually comes out in the form of delicate threads which hardens on exposure to air. Gradually the number of threads coming out of the caterpillar increase and they form a covering around the body called the cocoon.

Cocoon - The cocoon is a covering generally white in colour and is made up of closed weaving of the silk threads. In the cocoon the threads on the exterior are somewhat irregular while those present towards the inner region of the cocoon are regular in arrangement. It has been estimated that the length of a single continuous thread secreted by a pupa for the formation of cocoon is about 1000-2000 mts in length. The secretion of this thread requires about 3 days time. The arrangment of the thread in the cocoon is in a concentric manner. In order to arrange the thread in a concentric manner the pupa rotates its head at the rate of 65 times per minute constantly.

A fully formed cocoon has the pupa inside. Estimates have shown that 2500 cocoons can yield upto one pound of silk. In the weight of the cocoon (1.8-2.2 gms) the shell which is useful in extracting silk is only about 0.45 gms. The pupal stage takes about 10-12 days to complete the life cycle. At the end of this stage the pupa would have grown into an adult (imago) and it comes out of the cocoon by cutting it open.

Emergence of the adult moth *(imago)*- The pupa when it is inside the cocoon undergoes physiological and morphological changes called metamorphosis. The abdominal pseudo legs disappear and in its place two pairs of wings develop. The moth when it is inside the cocoon secretes an alkaline fluid which moistens one corner of the cocoon. At this point, the cocoon is soft and the moth cuts the threads to make a hole. It is through this the adult moth comes out of the cocoon.

CHAPTER 24

BIOCHEMICAL TESTS

TEST FOR PROTEINS

Object : To test the presence of proteins.

1. Biuret test

Requirements : Test, tube, protein solution (egg albumin distilled water), 40% NaoH, 1% copper sulphate solution

Procedure : Take a little quantity (4 ml) of protein solution in a test tube. Add about 2 ml of 40% NaOH and one drop of copper sulphate solution. Mix the ingredients thoroughly and note the colour.

Observation : A pale purple colour appears indicating the presence of protein.

***Reasons* :** In biuret reaction, an alkaline solution of biuret ($CONH_2$. NH. $CONH_2$) gives a colouration on reacting with the peptide bond in the protein.

Ninhydrin test

Requirements : Test tube, protein solution (egg albumin diluted with distilled water), Ninhydrin solution (0.1%)

Procedure : Take a little quantity (4 ml) of protein solution in a testube. Add to this 1 ml of freshly prepared 0.1% ninhydrin solution. Mix the contents thoroughly and boil for about two minutes. Allow the solution to cool.

Observation : A violet or purple colour develops

Reasons : The ninhydrin test is answered by proteins as well as free amino acids. A complex called Rhenmann's purple is formed. This is due to the condensation of two molecules of nin hydrin with one molecule of ammonia from amino acids. The result is the formation of purple or violet colour.

Test for carbohydrates

Object : To detect the presence of carbohydrates

Fehling's solution test

Requirements - Carbohydrate solution (glucose solution in distilled water); Fehling's solution A (dissolve 34.65 gm of copper sulphate in distilled water and make up to 500 ml); Fehling's solution B (dissolve 125 gm potassium hydroxide and 173 gm of Rochelle salt - potassium sodium tartarate in distilled water and make up to 500 ml); Test tube etc.

Procedure - To one ml of Fehling's solution , A, add one one ml of Fehling's solution B, and a few drops of test solution. Boil for few mintues and cool.

Observation- A yellow or brownish red precipitate is formed

Reasons - The blue alkaline cupric hydroxide present in the Fehling's solution, in the presence of reducing sugars gets reduced to yellow or red cupric oxide which gets precipitated. The formation of yellow or red precipitate indicates the presence of reducing sugars (carbohydrates)

Iodine Test

Requirements - Startch solution; Iodine solution (Add a few crystals of Iodine to 20 potassium iodide solution till the colour becomes deep yellow); testubes etc.

Procedure - Add a few drops of Iodine solution to 1 ml of starch solution and note the colour change

Observation - A deep blue colour develops

Reasons - The blue colour is due to the formation of starch iodine complex, indicating the presence of starch.

Test for Fats (Acrolein test)

Requirements - Any oil-olive, castor etc; potassium bisulphate, test tubes etc.

Procedure - In a clean test tube take about 1 gram of potassium bisulphate and add 4-5 drops of oil. Stir thoroughly and heat. Acrolein is formed due to digestion of fat. A pungent smell also can be felt.

Reasons - Glycerol of fats gets degraded to acrolein with the release of water.

HUMAN URINE ANALYSIS FOR SUGAR

(i) Benedict's reagent test - Take about 5 ml of Benedict's qualitative reagent in a clean test tube and add to it about 5-10 drops of urine smaple. Boil for two minutes and allow it to cool. Formation of red, yellow or green coloured precipitate indicates the presence of sugars.

(ii) Fehling's test - Mix one ml each of Fehling's solution. A and B (for preparation of Fehling's solution-refer-test for carbohydrates) in a test tube. Add to this mixture about 5ml of distilled water and boil. Add a small amount of urine sample and boil again. Development of red or yellow precipitate indicates the presence of sugars.

BLOOD GROUP IDENTIFICATION TEST

In human beings there are four blood groups - A, B, AB and O which are differentiated by the antigenic reactions.

Requirements - Anti A serum, Anti B serum, few drops of fresh blood, clean glass slides, marking pencils etc.

Procedure- Take two clean glass slides. Mark on one slide - Anti A serum and on the other Anti B serum. Now prick the finger tip with the help of a sterilized needle and place a drop of blood each on both the slides. Now add a drop of Anti A serum to the slide marked Anti A serum and a drop of Anti B serum to the slide marked Anti B serum. Mix the ingredients thoroughly with a clean needle, and observe. The following results show to which group the blood belongs.

(i) If there is clumping (agglutination where RBCs aggregated) only in Anti A serum and no clumping in Anti B serum the blood group is A.

(ii) If there is no clumping in Anti A serum and clumping in Anti B serum, the blood group is B.

(iii) If there is clumping in both Anti A serum and Anti B serum, the blood group is AB.

(iv) If there is no clumping in both Anti A serum and Anti B serum, the blood group is O.

ANALYSIS OF MILK - Lactometer test for water content and test for total proteins (precipitate test)

Lactometer test for water content

This instrument is used to test the density of milk; if the density is lower it indicates higher content of water.

Lactometer is a glass tube with one end having a bulb and the other end is a narrow graduated tube. The bulb consists of weights (lead pellets or mercury) which will make the lactometer stand upright when immersed in a liquid. The graduations on the tube range from 25° to 35° which indicates the specific gravity of the liquid.

Specific gravity of liquids- Water is having a specific gravity of 1000 and that of milk (cow's) varies between 1028 to 1034. In toned milk the specific gravity is low whereas in skimmed milk it is high.

Lactometer is dipped in the milk to be tested (for its water content) and the level at which it floats in milk indicates its specific gravity. If the level is between 25° to 35°; then it can be assumed that no water has been added to the milk. If the level falls below 25° it is an indication of excess water ie. milk is diluted.

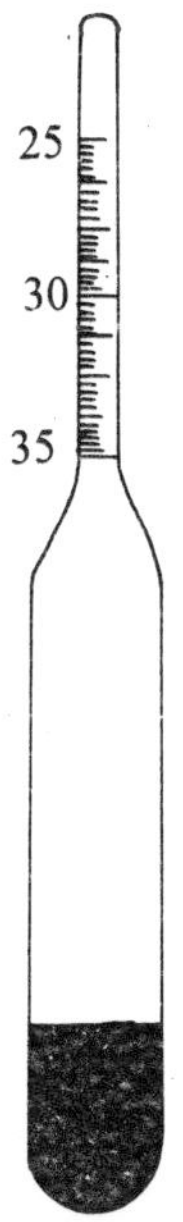

Fig. 24.1 Lactometer

Test for total proteins (precipitate test)

Milk contains proteins like caesin and albumin which have to be separated before they are identified. In order to separate the proteins from the rest of the milk constituents; dilute the milk with about 2-3 times water and add 10-15 drops of acetic acid (25ml of milk+75ml of water). Warm the mixture gently on a low flame of spirit lamp. A precipitate appears in the liquid. Separate the precipitate and use it to test for proteins; this consists total milk proteins. Test for proteins as follows.

(i) **Biuret reaction** - Add about 2ml of 10% NaOH to the precipitate (taken in a test tube), mix thoroughly . Add two drops of 0.1% copper sulphate solution. A violet or pink colour is produced indicating the presence of proteins.

(ii) **Xanthoproteic reaction** - Add to the precipitate (taken in a test tube) 1 ml of concentrated nitric acid. Carefully boil and cool the contents. After cooling add ammonium hydroxide (NH_4OH) or sodium hydroxide (NaOH).

On adding nitric acid yellow colour is formed which changes to orange on the addition of alkali like NH_4OH or NaOH. This indicates the presence of aromatic amino acids in the protein.

Visit to a nearby dairy/ poultry/sericulture : Farm to get a first hand working knowledge or to a nearby primary health centre/ clinic/ nursing home to get a first hand knowledge in social medicine, AIDS.

The vistis to above mentioned places will help the students to understand the working of these better. They can also collect useful data on the various diseases, their symptoms, cure etc.